The Physics of Reservoir Fluids:
Discovery Through Downhole Fluid Analysis

Oliver C. Mullins

Schlumberger
225 Schlumberger Drive
Sugar Land, Texas 77478
www.slb.com

Copyright © 2008 Schlumberger. All rights reserved.

No part of this book may be reproduced, stored in a retrieval system, or transcribed in any form or by any means, electronic or mechanical, including photocopying and recording, without the prior written permission of the publisher. While the information presented herein is believed to be accurate, it is provided "as is" without express or implied warranty.

Specifications are current at the time of printing.

08-FE-023

An asterisk (*) is used throughout this document to denote a mark of Schlumberger. Other company, product, and service names are the properties of their respective owners.

Downhole fluid analysis, once only a novel idea, is now an oil industry standard methodology that generates essential and abundant information about the primary and elusive object of our study, the reservoir. This dramatic evolution in both science and technology is founded on the spirit, creativity, and tireless effort of a broad and growing community of scientists and engineers. The author gratefully acknowledges the fundamental contributions of these progressive and dynamic professionals who hail from all levels of the oil industry and whose only collective limitation is the fortunate inability to recognize conceptual boundaries.

The author also acknowledges the enduring and unconditional support of his family which has provided the foundation for his lifelong journey into the exquisite beauty of science and to the profound satisfaction of its application. In particular, the author acknowledges his father, William Wilson Mullins (now deceased), former University Professor of Applied Science, Carnegie-Mellon University. Professor Mullins's scientific legacy encompasses inclusive and encouraging leadership coupled with rigorous and unrelenting scientific excellence; any successful fulfillment by the author of these lofty goals was borne by lessons learned at his father's side.

Contents

Chapter 1:
The Distribution of Reservoir Fluids and Their Characterization by DFA

Overview 1
The complexity of reservoir fluids 3
Equilibrium distribution of hydrocarbons 5
DFA and equilibrium distributions of hydrocarbons 10
    DFA density measurements 14
Asphaltenes and equilibrium fluid distributions 15
Nonequilibrium distributions of hydrocarbons 24
    Transients across geologic time 24
        Current reservoir charging: Hydrocarbons 28
        Current reservoir charging: Nonhydrocarbons 32
        Reservoir charge history 35
Compartments 43
    Compartments: Problem 1 44
    Compartments: Problem 2 45
    Compartments: Problem 3 46
Production and miscible injection 55
    Flow assurance and compositional variation: Asphaltenes 55
    pH and water chemistry 58
DFA workflows 60
    The new DFA workflow 61
Senior technologists 62
Conclusions 63
References 64

Chapter 2:
The Photophysics of Reservoir Fluids: The Scientific Foundation of Optical DFA Measurements

Introduction 71
    Oil sample acquisition in open hole 71
    MDT Modular Formation Dynamics Tester 72
Vis-NIR spectroscopy of crude oil and water 74
    Fluid identification 74
    GOR, the first formal DFA product 76

Oil chemistry 82
    Analytical chemistry 84
    Molecular spectroscopy 86
        Molecular energy levels 86
        Basic optical principles in spectroscopy 90
        Scattering 93
Crude oil color 100
    Temperature dependence of crude oil coloration 113
    Quicksilver Probe focused fluid acquisition and asphaltenes 114
Hydrocarbon compositional analysis 116
Index of refraction and gas detection 129
    Hydrocarbon phase transitions: Gas 129
Fluorescence of crude oils 135
    Downhole fluorescence 135
    The science of crude oil fluorescence 142
Transition zone characterization and downhole pH measurement 153
Chain of custody 156
DFA fluid profiling: A quasi-continuous downhole fluid log 158
    Cost efficiency: Toward a continuous fluid log 158
Petroleomics, the future of hydrocarbon analysis 159
Conclusions 166
References 167

Index 177

Preface

Downhole fluid analysis (DFA) is relatively new; nevertheless, DFA is now an essential and growing product line in the oil industry. Indeed, it is uncommon to have a new technology become such a keystone in formation evaluation. Upon this occurrence, special consideration for the corresponding dissemination of a definitive overview is mandated. DFA is expanding to include many fundamentally different analytic methods for fluid evaluation. Nevertheless, optical methods are the foundation and focal point of DFA. As such, it is desirable to have a comprehensive description of the new technology itself, as well as a full explanation as to why the application of this technology is so indispensable. In particular, DFA relies on bulk optical spectroscopy for chemical analysis while (upstream) PVT laboratories rely on separation science such as gas chromatography for chemical analysis. Consequently, there is no prior compendium relating optical spectroscopy to the objectives of the oil industry in the upstream setting.

Quintessential DFA implications relate to the reservoir, the object that remains the largest technical uncertainty in the oil industry. Indeed, recent realizations regarding complexities of reservoirs and their contained fluids have been driven in part by DFA application. In particular, fluid compositional variations in reservoirs, compartmentalization, and connectivity have been established by DFA in new ways. The end users of DFA data will be those who have responsibility for the reservoir, such as the asset managers. This exceedingly important role might or might not be occupied by technologists. The primary technologists on the asset team are often reservoir engineers. If there are chemists on the reservoir technology team, often they are specialized in disciplines different from optical spectroscopy. These considerations have shaped the focus in this book.

The first chapter, "The Distribution of Reservoir Fluids and Their Characterization by DFA," is written to be informative if not provocative, stimulating, and readable for persons of all backgrounds, not just for expert chemists or DFA practitioners. Indeed, I believe that this chapter will be of interest to a very broad audience of managers and technologists alike in the upstream oil industry. In contrast, the chapter titled "The Photophysics of Reservoir Fluids: The Scientific Foundation of Optical DFA Measurements" is meant to establish in broad context the scientific validity of DFA measurements—a required stipulation of the author. In addition, this second chapter is meant to provide to the inquisitive the scientific underpinnings of DFA. That is, it will be of use to a subset of the audience of the first chapter. For example, the many supremely skilled reservoir engineers who apply DFA for a variety of objectives will find utility in the scientific principles delineated in the second chapter, thereby enabling greater power and capability in the application of DFA to reservoir concerns.

1 CHAPTER ONE
The Distribution of Reservoir Fluids and Their Characterization by DFA

The Physics of Reservoir Fluids

Overview

Significant complexities of oil reservoirs, both their architecture and their contained fluids, are increasingly realized to be commonplace and are now expected as the norm. These insights override previously prevailing, yet unsupportable, expectations regarding reservoirs as "one homogeneous hydrocarbon in one giant tank." However, the combined technologies of reservoir reconnaissance have not proffered moderately priced, effective delineation of reservoir complexities. Consequently, these complexities have gone unrecognized until after enormous expenditure in facilities and production strategies that are subsequently proved incommensurate with previous expectations. There is no magical cure for this condition.

Nevertheless, huge strides have recently been made that greatly improve reservoir reconnaissance with sufficient efficacy to become an industry standard in short order—this in an industry with an undeserved reputation for being slow to adapt. The foundation of the recent technological and indeed scientific breakthroughs is the in situ characterization of reservoir fluids within individual wells and collectively in many wells within a reservoir context. This technology is subsumed under the banner of downhole fluid analysis (DFA), and the resulting evaluation of reservoir fluids is described as reservoir Fluid Profiling* characterization of reservoir fluid properties and quantification of their variation. Fluid Profiling evaluation by DFA necessarily delineates reservoir fluid complexities and, moreover, naturally addresses key attributes of reservoir architecture.

Prior to DFA, there was no efficient means to perform Fluid Profiling evaluation. Consequently, simple programs of sample acquisition and analysis prevailed along with inherent and concomitant assumptions of the simplicity of reservoir fluids. These assumptions caused repeated disasters for operating companies, leading to an industry poised for a change. Indeed, inefficiencies in any technological markets are no longer tolerated if emergent new solutions become available. DFA represents a paradigm shift in well logging and in reservoir evaluation. As is so often the case, this new technology has spurred new scientific understanding of oil reservoirs, which is driving the need for more expansive DFA measurement suites. DFA is a totally new fluid data stream outside traditional workflows, and thus it establishes opportunities and challenges for creativity. DFA is the linchpin of all fluid analyses for addressing the largest uncertainty in the oil industry today—the reservoir.

The acquisition and analysis of openhole wireline samples are increasingly important for reservoir delineation. DFA has become an indispensable tool for addressing reservoir complexities in a cost-effective manner through the real-time measurement of fluid properties under reservoir conditions. For simple hydrocarbon (and water) columns, DFA provides the means to monitor the validity of sample acquisition. In complex fluid columns, DFA is of penultimate importance because "blind" and copious sample acquisition is prohibitively expensive and does not necessarily obtain representative results. It is not known a priori whether fluid simplicity prevails, thus using DFA in real time to delineate fluid complexities and confirm the necessity of fluid measurements (both DFA and laboratory) is technically sound. Moreover, additional MDT* Modular Formation Dynamics Tester sampling stations for DFA while the tool is still in the well represent a marginal cost increase. Consequently, DFA has greatly expanded the investigative capabilities available to the industry for revealing fluid complexities and enabled new discovery of the physics controlling reservoir fluid variation. Key to using DFA to ferret out the manifold conditions that produce fluid complexities is the collaborative effort of operating companies and Schlumberger.

Indeed, the understanding of reservoir fluids is rapidly evolving and mandates vigilance to remain up-to-date. More importantly, the symbiotic coevolution of DFA and the understanding of reservoir fluids challenges those in the fold to lead new developments, and opportunities abound. As of this writing, Schlumberger has produced more than 70 publications in peer-reviewed scientific journals that describe the physics behind DFA measurements. A similar number of applications papers with manifold operating companies as coauthors have been published in oilfield journals, delineating reservoir complexities and their investigation with DFA. The growth of DFA literature is substantial as the new discipline is developed. Whereas these publications focus on the specifics of particular DFA applications, this chapter provides an overview of the complexities of reservoir fluids, the physics behind the complexities, and their interface with DFA.

At Schlumberger, we take a practical as well as a scientific approach; this document reflects both tacks. Our goal is to understand the major features of the reservoir fluid profile within a wireline context, thereby identifying the key governing physics that controls these fluid properties in a timely manner. Our focus is to reveal as much as possible within a cost-benefit framework about the reservoir and contained fluids prior to production with its associated large capital expenditure. To achieve that goal, Schlumberger has led in launching DFA on wireline and introducing the enabling DFA tools: LFA* Live Fluid Analyzer, CFA* Composition Fluid Analyzer, and InSitu pH* Reservoir Fluid pH Sensor; this last tool brings water chemistry into DFA focus. The recently launched InSitu Family* quantitative measurement of fluid properties at reservoir conditions expands the representative, real-time data foundation for DFA. This vision will naturally extend into formation evaluation while drilling when corresponding tools are deployed. The objective is to provide critical data early in the process, independent of conveyance; this is when the albeit limited data is of extreme value. As such, our technical perspective deviates from traditional scientific studies, for which perhaps more complete datasets are acquired but no consideration is made regarding the decay of the economic value with time-delayed delivery of results. Nevertheless, our perspective is wholly scientific while incorporating this economic vantage.

The Schlumberger vision of expanding DFA beyond the traditional constraints of sample acquisition into a new methodology for investigating the reservoir is being fulfilled. DFA jobs with more than 30 stations are providing timely insight in single wells for a wide variety of well depths and complexities. Indeed such extensive DFA jobs have been run in different market segments from very high cost structure to low cost structure. The vision is correct and proven, and optimal DFA application is now the target.

The complexity of reservoir fluids

Figure 1 shows 24 dead-oil samples obtained from a single column in a deepwater Gulf of Mexico field (courtesy Hani Elshahawi, Shell International E&P). The height of the oil column is several hundred meters, over which the reservoir crude oils exhibit a huge compositional gradient. However, unless DFA is employed, it is not the norm to acquire samples from 24 or more different stations. Without targeted sampling and analysis such complexity remains hidden, yet can cause large, expensive production problems.

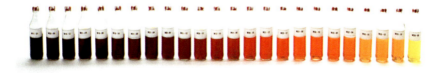

Figure 1. A series of dead crude oil liquids from a single oil column visually affirms a large compositional gradient for the reservoir fluids. Photograph courtesy of Hani Elshahawi, Shell International E&P.

The large gradient of the dead oils in Fig. 1 is visually evident. For live oils in the reservoir, the GOR is very high for the lightly colored samples at the top of the column (right end) and very small for the heaviest samples at the bottom of the column (left end). Consequently, the visual contrast would be substantially increased if these were live-oil samples. What causes these gradients in the GOR and color (the latter related to the asphaltene content)? They can be governed by different physics, and it is important to understand their origin.

In the past, fluids were analyzed by organic geochemists in large part to analyze biomarkers to test for compositional variations and for compartments.[1] However, these methods, although useful, suffer from significant shortcomings that render their output of somewhat limited utility. First, the analysis is not in real time during downhole sample acquisition; the results arrive very late for some key decision making. Second, the biomarkers are present in crude oils only in very low concentrations, and the factors that influence the biomarkers may not have much relevance for the bulk of the crude oil (Wilhelms and Larter, 2004). Third, biomarkers vary in concentration in different crude oils by 5 orders of magnitude (Wilhelms and Larter, 2004). An immature charge can bias conventional biomarker analysis because of the high concentration of biomarkers in immature charges; biomarker analysis can totally miss a significant secondary, mature charge. Wilhelms and Larter (2004) state that it is much better to analyze light end to heavier end ratios (such as GOR) rather than conduct biomarker analysis because the former much more accurately reflects the complete reservoir charge. Their publication essentially anticipates DFA methods for characterizing fluids. Thus, DFA is recognized as founded on a sound basis. The challenge to the industry is to incorporate the new DFA data stream. Forward thinking is required to exploit new kinds of measurements, and the rapid uptake of DFA by the petroleum industry proves this industry is rather agile, contrary to conventional reputation.

Consider the related concern of understanding the reservoir rock. Imagine the following scenario: a sidewall coring tool collects a single sidewall core. The core is sent to a laboratory where a large, impressive report is written about the elemental composition, mineralogical description, and petrophysical characterization of the core. The declaration is made that the rock in the reservoir is now understood. If such a workflow were advocated for reservoir rock evaluation, the advocate would be summarily fired. It is understood by all that the reservoir rock can vary in a centimeter-length scale. Yet, in the past, a similar workflow was advocated to get and analyze a sample of "the oil" in the reservoir. But we are not in the rock business, we are in the oil business. It is now time to give proper respect to oil in the oil industry and to understand it too can exhibit large spatial variations. DFA is the enabler to shed light on these fluid variations that have remained concealed for so long.

---

[1] We define compartments as fluid-containing reservoir bodies that must be penetrated by a well to drain.

Equilibrium distribution of hydrocarbons

Høier and Whitson (2001) among others have written a series of papers illustrating conditions that can give rise to large compositional variations in oils. We refer to these equilibrium gradients as Høier-Whitson gradients, but many workers have contributed to this understanding. Figure 2 shows that when reservoir conditions are near the critical point, or reservoir pressure is near saturation pressure, or both conditions hold, then large compositional variations are expected.

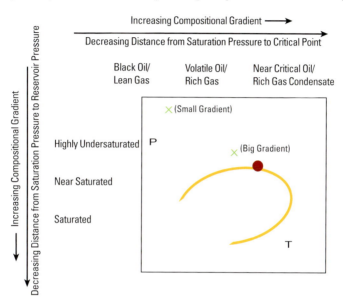

Figure 2. Near-critical and near-saturation fluids exhibit large compositional gradients. After Høier (1997).

The inset pressure-temperature plot illustrates the corresponding phase diagram. For liquids, this figure primarily addresses compressibility: highly compressible fluids exhibit large gradients in GOR whereas incompressible liquids exhibit small GOR gradients. Why? If a fluid is compressible, then the hydrostatic head pressure of the oil column squeezes the bottom of the fluid column to a higher density. The density gradient creates the thermodynamic drive for the compositional gradient. The system reacts to move low-density components such as (dissolved) methane out of the bottom of the column and into the top of the column, where the low-density fluid is. This is a manifestation of Le Chatelier's principle, which states that an equilibrium that is stressed (for example, by a hydrostatic gradient on a compressible fluid) responds to relieve the stress (by preferentially removing low-density components from the bottom of the column). For liquids, a large gas fraction increases compressibility. Fluids under very high pressure (thus highly compressed) are less compressible, and such fluids are highly undersaturated. In contrast, for an incompressible fluid, hydrostatic head pressure has no effect. Because the column is incompressible, there is no "strain" induced by the stress, thus no GOR gradient occurs.

For gases, compositional grading still depends on compressibility. In addition, the extent of compositional grading depends on the density contrast of the constituents. The greater the contrast, the greater the compositional grading as gleaned in Fig. 2. For rich gases with a large column height, Le Chatelier's principle argues that larger, denser alkanes such as pentane should settle near the high-density bottom of the column while smaller, lower density components such as methane should rise to the top of the column. At the limit of pure methane (presuming one isotope ratio), there can be no compositional grading.

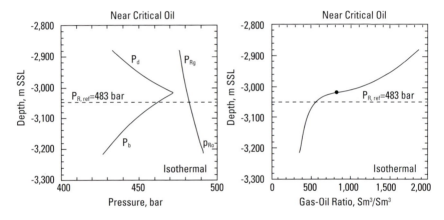

Figure 3. This near-critical fluid has large compositional gradients. The calculations are for an isothermal gravitational field assuming gravitational chemical equilibrium with a slightly undersaturated GOC. From Whitson and Belery (1994). © 1994 Society of Petroleum Engineers.

Figure 3 is an example of a compositional gradient. Near-critical fluids have a large variation in GOR and a large variation of the saturation pressure with depth. This is the expected behavior for a near-critical column in equilibrium. Figure 3 also shows that the minimum difference between reservoir pressure and saturation pressure is at the GOC. Sampling at or close to the GOC can result in phase transitions, so it should always proceed with caution.

Laboratory experimental confirmation of the Høier-Whitson predictions is found in a highly cited publication by Ratulowski, Fuex, Westrich, *et al.* (2003). A centrifuge tube apparatus was used to create large g-forces on prepared hydrocarbon mixtures and on live Bullwinkle reservoir crude oils from deepwater Gulf of Mexico (Fig. 4).

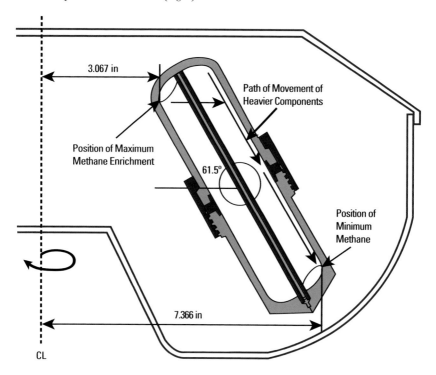

Figure 4. A centrifuge was used to create large g-forces on hydrocarbon mixtures and crude oils. The geometry during spinning is shown; the centrifuge tube is vertical at rest. From Ratulowski, Fuex, Westrich, *et al.* (2003). © 2003 Society of Petroleum Engineers.

The samples were spun for up to 23 days in the centrifuge apparatus to achieve equilibrium, producing large GOR gradients. In addition to the large compositional gradient prepared with the centrifuge, a good match was obtained for the centrifuge data and equation-of-state (EOS) modeling (Fig. 5).

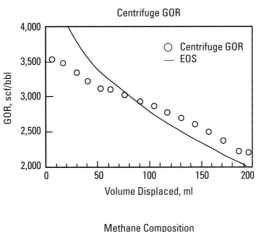

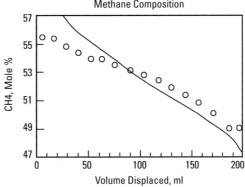

Figure 5. A good match was obtained for laboratory data and EOS modeling for GOR and methane compositional variations for the sample subjected to 2,500 g's for 7 days. The prepared hydrocarbon mixture is methane, n-pentane, and 1-methyl naphthalene in a mole ratio of 50:30:20. From Ratulowski, Fuex, Westrich, et al. (2003). © 2003 Society of Petroleum Engineers.

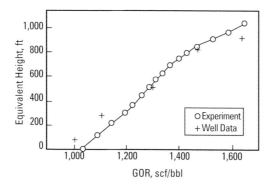

Figure 6. Centrifugation of the Bullwinkle live crude oil matched the field data, confirming the existence of a large compositional gradient in GOR. From Ratulowski, Fuex, Westrich, *et al.* (2003). © 2003 Society of Petroleum Engineers.

Reservoir fluids were also centrifuged, and the centrifuge data matched the field data reasonably well (Fig. 6). This crude oil does not have a large GOR; it is the large-GOR oils that are compressible and tend to exhibit larger gradients. However, the equilibrium in GOR does not mean that there is an equilibrium in the asphaltene content. That is, GOR can be in equilibrium while the "color" measured by DFA can be out of equilibrium. More to the point, any unmeasured component cannot be assumed to be in equilibrium—this must be established.

DFA and equilibrium distributions of hydrocarbons

Fittingly, the first example of DFA finding a Høier-Whitson column (Fig. 7) was in Norway (Fujisawa, Betancourt, Mullins, et al., 2004). The Tyrihans MDT job conducted by Michael O'Keefe and Brett Mitchell of Schlumberger is canonical in that it was the first use of the CFA Composition Fluid Analyzer in an exploration well and a huge gradient was found but was not seen in the density profile of the fluid; consequently, this job generated significant surprise to put it mildly. Note that the pressure data is not in conflict with the compositional data; rather, the pressure gradient data is not sufficiently sensitive to observe the compositional gradient.

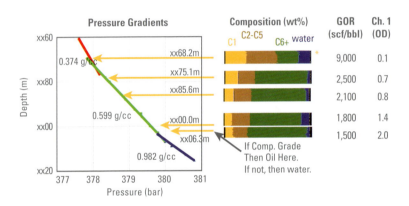

Figure 7. The first deployment of the CFA Composition Fluid Analyzer in an exploration well revealed a large GOR compositional gradient. The large number of DFA stations in 40 m was justified owing to the clear evidence in real time that a large gradient exists. The compositional gradient mandated a curved, not straight, pressure gradient line, lowering the estimated OWC. This was confirmed by sampling oil, not water at X,X06.3 m. The gas cap was shown to be a retrograde dew by DFA methods, again confirming the DFA-measured gradient. From Fujisawa, Betancourt, Mullins, et al. (2004). © 2004 Society of Petroleum Engineers.

The GOR gradient was identified during the job as successive DFA stations were added, motivated by the variation of GOR that was being measured. The technologists involved, Michael O'Keefe of Schlumberger and Kåre Otto Eriksen of Statoil, were instrumental in modifying the original MDT job plan to exploit the new DFA measurements to understand the reservoir fluids. The large GOR gradient found with added DFA stations implied that the pressure gradient is curved; consequently, the gas/oil contact (GOC) and oil/water contact (OWC) move apart (Fig. 7). This was checked in real time by running a DFA station at the bottom of the column; a straight oil-gradient line would predict water whereas a curved oil-gradient line would predict oil. Oil was found, convincing the client that the GOR gradient was real. Further confirmation of the large gradient was made by running DFA in the gas cap to check whether it was a retrograde dew, a finding which would further reinforce that the oil was a near-critical fluid. By breaking the gas cap into two phases, confirmation was obtained that the gas was a retrograde dew (Betancourt, Fujisawa, Mullins, *et al.*, 2004). Because the oil was near critical it could exhibit large gradients in GOR, as predicted by Høier-Whitson. The fact that the contacts diverged allowed the client to properly book more reserves, which is obviously a big motivation to understand reservoir fluids. The team of technologists was key in assessing in real time the implications of the DFA data and developing further DFA tests, again in real time, to validate the findings. The increased costs of the MDT job were more than offset by better understanding of the reservoir and booking more reservoirs.

An important point is that the client did not sample the gas after breaking the gas (retrograde one-phase dew) into two phases. This is because a two-phase sample is nonrepresentative. Even if the sampling drawdown is reduced and the sample becomes single phase, there could be residual problems from previously breaking the reservoir fluid two-phase; for example, with mini–condensate banking. The DFA station in the gas was used to determine the fluid identity as a retrograde dew in large part to confirm the GOR gradient but not necessarily for sampling.

Subsequent to this work, Dubost, Carnegie, Mullins, *et al.* (2007) showed that the GOR gradient in this column can be treated with standard EOS modeling. Figure 8 shows the comparison of the EOS model and the CFA data. This important result confirms the basis for the DFA observations: the oil column is most likely in an equilibrium condition. One lesson from this is that all near-critical condensates can exhibit similar gradients and should be similarly investigated using DFA.

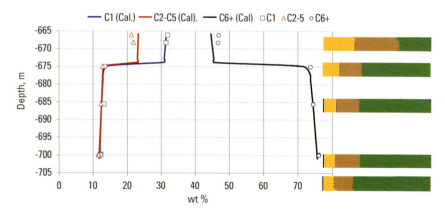

Figure 8. The modeled and measured mass fractions compare well for the compositional variation in Fig. 7. The modeling assumes the fluid column is in equilibrium. From Dubost, Carnegie, Mullins, *et al.* (2007). © 2007 Society of Petroleum Engineers.

In a recent development, this Tyrihans reservoir was again penetrated by a well. EOS predictions of the expected fluids were developed by Julian Zuo of Schlumberger and compared with LFA and CFA log data acquired during the job. The comparison, conducted by Adriaan Gisolf of Schlumberger and Rolf Magne Pettersen and Steve Williams of StatoilHydro, confirmed that the encountered fluid was mostly as predicted from the EOS modeling and from the previous DFA data. Indeed, for one DFA station, the data did not match thus an additional DFA station was added nearby and again predictions matched observations. The conclusion was reached that sample acquisition in the aberrant DFA station was probably not representative. Other implications of the DFA data are currently in assessment. Again the technology team across corporate boundaries reached a consensus and optimal MDT use was the outcome, benefiting both the service company and the operating company.

Herein lies the power of DFA toward the vision of bringing proper predictive science to the study of the reservoir. In this latest Tyrihans job, DFA log predictions were made based on a presumed accurate static model of the reservoir. There is too much focus in reservoir engineering on simplified dynamic models without testing the more accurate *static* models. To test the static geologic and fluid models, the predicted DFA log data was compared with the acquired DFA log data. The major reservoir attributes are being tested with data that is validated, all in real time (the corresponding interpretation is not yet released). When some concerns arose with certain DFA data, this was addressed in real time by senior technologists in both the operating company and the service company, leading to a satisfactory resolution. Because the analysis was done during log acquisition, the logging program could be modified to *measure* the nature of the discrepancy. This is much preferred over common current practices in reservoir engineering, in which error is determined on a computer by repetitively running simplified dynamic models. There is no other experimental discipline known to humankind in which error is determined solely on a computer. DFA offers the potential to cast reservoir engineering into the tried and proven scientific precepts that make the scientific method so powerful. Another case study involving asphaltenes is explored in "Asphaltenes and equilibrium fluid distributions" in this chapter.

A question arises as to why this large compositional gradient is not registered more clearly in the density profile, that is, in the wireline log pressure gradient. Other treatises have expounded on expected errors in pressure gradient measurements. A measurement that sheds light on this issue is that density ρ is an integral quantity, the sum of the masses of all contributing chemical moieties in a volume ($\rho = (\Sigma_i m_i)/V$), and is thus rather insensitive to changes in the small-scale structure of the substance. In contrast, compressibility ($\beta = (-1/V)(\delta V/\delta p)$) is a differential quantity and is thus much more sensitive to changes in the small-scale structure of the substance. Figure 9 shows this for high-Q ultrasonics measurements[2] of a standard surfactant (sodium

---

[2] *The* Q*, or quality, factor of a resonator is defined as the resonance center frequency divided by the width of the resonance (full-width half-maximum). A very large value of* Q *means that the resonance lines are very narrow so that the frequency and frequency shifts can be measured with great accuracy. Equivalently, the* Q *of a resonator is the ratio of the energy stored divided by the energy dissipated per radian of the oscillation of the (acoustic) field. A high* Q *means that energy loss is minimal.*

dodecyl sulfate [SDS]) going from a true molecular solution in water to a surfactant micelle upon an increase in SDS concentration (Andreatta, Bostrom, and Mullins, 2005, 2007). Soap, a classic surfactant, cleans by dissolving oil substances in the nonpolar interior of micelles. The surfactant SDS is in many shampoos, soaps, and other personal care products.

The concentration of micelle formation can be obtained by measuring the change in solution properties versus the concentration of the surfactant. Surfactants by definition have an affinity for the interface. The initial surfactant added to water (or other solvent) preferentially loads onto the water-air surface. At the concentration when the surface is fully loaded (the critical micelle concentration, or CMC), additional surfactant forms micelles in the bulk solvent. Both the change in density and ultrasonic velocity versus SDS surfactant concentration are plotted in Fig. 9. These measurements were made with a very high-Q system (~10,000), enabling the speed of sound to be measured with a precision of a few parts per million (Andreatta, Bostrom, and Mullins, 2005, 2007). Sonic velocity is given by $u=\sqrt{1/\beta\rho}$.

In Fig. 9, the enormous change of the slope of ultrasonic velocity and the barely detectable change of the slope of density reflect the insensitivity of the integral quantity in comparison to the differential quantity. The aggregation threshold for asphaltenes in toluene was established for the first time with these high-Q ultrasonic measurements (Andreatta, Bostrom, and Mullins, 2005, 2007) and have been confirmed by many techniques. DFA measurements are essentially differential—there is direct interrogation of particular analytes in the sample, as opposed to measuring an integral bulk property and inferring compositional change, as is performed in pressure gradient analysis.

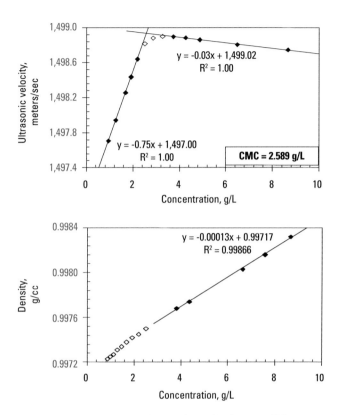

Figure 9. Plots of measurements of ultrasonic velocity and mass density versus SDS concentration in water show an insensitivity of density, an integral quantity, and excellent sensitivity of sonic velocity to micelle formation in the fluid. The acoustic velocity measurement is a differential measurement similar to DFA compositional measurements, which detect a particular analyte. CMC = critical micelle concentration. From Andreatta, Bostrom, and Mullins (2005, 2007). © 2005 American Chemical Society.

DFA density measurements

Nevertheless, the direct measurement of fluid density is a valuable addition to the DFA suite of measurements. Density impacts both primary MDT measurements of fluid properties and pressure. Fluid density links these two independent measurements into a single framework, enabling extensive consistency-checking of the two different but interconnected physics of fluids and pressure. Pressure gradients are used to determine fluid density in the formation. This fluid density is for the formation fluid, not the filtrate. Filtrate invasion can induce a constant pressure offset because of capillary pressure effects. However, pressure gradients in transition zones can become quite complex and even show negative slopes (Carnegie, 2006). The DFA density measurement can immediately be compared with density derived from pressure gradients. For example, once a compositional gradient is identified, it can be compared with a variable pressure gradient (Elshahawi, Venkataramanan, McKinney, et al., 2006). The measurement of density makes the comparison more direct.

An additional check for light hydrocarbons and gases is that the measured density should be equal to the density determined by the sum of the components as determined by near-infrared (NIR) spectroscopy (Mullins, Elshahawi, Hashem, et al., 2005). For the heavier hydrocarbons for which density is not derived from the NIR spectra, consistency checking is still possible, for instance by comparing fluid density variations with relative asphaltene content and GOR obtained from spectroscopy. The overriding message is that there is limited data available regarding subsurface fluids, and once the MDT tool comes out of the well, there is often little chance of another trip into the well. Redundancy and consistency checking are vital for validating acquired data. The observation of inconsistencies in real time enables the acquisition of more DFA data to resolve the differences, thereby greatly improving the efficiency of MDT measurement and sampling runs.

Asphaltenes and equilibrium fluid distributions

The Høier-Whitson analyses account for the GOR and liquid-gas composition of equilibrium fluids as measured by DFA. DFA also measures oil color, which is directly proportional to asphaltene concentration, at least for the oils within a given field (Mullins, Betancourt, Cribbs, et al., 2007). So what is the distribution of asphaltenes in an equilibrium case? The Høier-Whitson and other EOS models do not properly address asphaltenes from a first principles approach. The primary reason for this is that the molecular and colloidal description of asphaltenes has been unknown in laboratory measurements, let alone for a live fluid in the reservoir. But this has changed.

DFA provided the industry's first look at how to analyze asphaltenes in an equilibrium setting of a live black oil with data from the Tahiti field, deepwater Gulf of Mexico (Fig. 10) (Mullins, Betancourt, Cribbs, et al., 2007). Indeed the stage for this was set by the tremendous advances in asphaltene science in recent years (Mullins, Sheu, Hammami, et al., 2007). Figure 11 shows the sands in the wells, organic geochemistry data, and pressure gradient data for Tahiti field (Betancourt, Dubost, Mullins, et al., 2007). The geochemistry (spider plot of gas chromatography data) and pressure data show that the M21A and M21B sands are not communicating. No connectivity problem is shown by geochemistry or pressure within each sand. In addition, the reservoir fluids are highly undersaturated black oils, very far from the critical point. Høier-Whitson indicates very little compositional variation should be observed for GOR if the reservoir fluid is in equilibrium. The Tahiti field is a tilted sheet reservoir with a large horizontal permeability (k_h = 600 mD) and large vertical permeability (k_v/k_h ~ 0.6) from core.

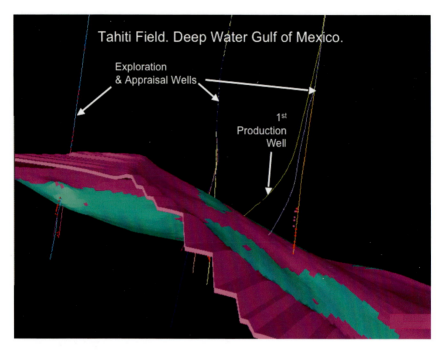

Figure 10. The upper and lower horizons of the Tahiti reservoir are shown along with several wells. From Mullins, Betancourt, Cribbs, *et al.* (2007) and Betancourt, Dubost, Mullins, *et al.* (2007). © 2007 American Chemical Society.

As pointed out by England, Muggeridge, Clifford, *et al.* (1995), with convection, fluid equilibration times in reservoirs can be 1 million years. However, if diffusion must take place across an anticline, the equilibration times can take 100 million years (England, 1990). Jefferson Creek of Chevron (pers. comm., 2005) notes that the Tahiti field parameters are conducive to establishing equilibrium. High-permeability enables convection, and a tilted sheet is a simple structure. Indeed the GOR variation is small, as predicted by the Høier-Whitson heuristics—and this condition is consistent with the reservoir fluids being in equilibrium. However, the reservoir is against a salt canopy, and the number-one risk factor is compartmentalization resulting from faulting induced by salt buoyancy.

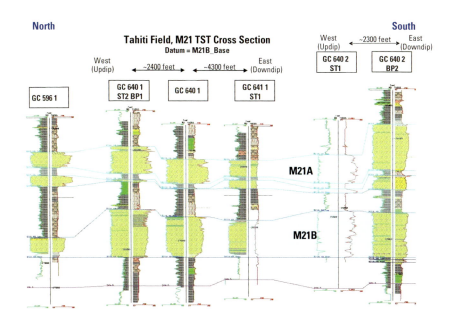
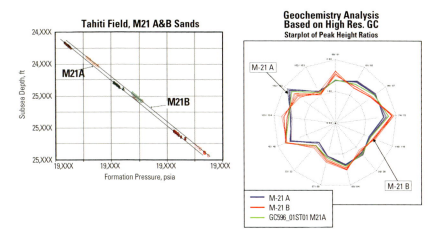

Figure 11. Tahiti field data for pressure and geochemistry show that the M21A and M21B sands are not connected to each other. Within each sand, these data do not show compartmentalization. From Betancourt, Dubost, Mullins, et al. (2007). © 2007 Society of Petroleum Engineers.

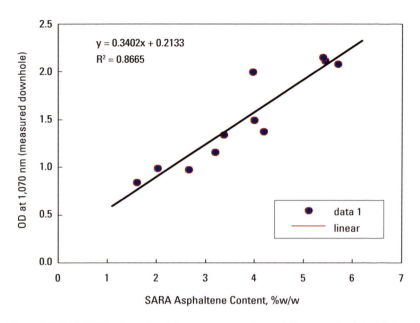

Figure 12. Tahiti field DFA color and asphaltene content correlate well. The uncertainty is largely from the laboratory determination of asphaltene content. The one outlier (2 OD, 4%) was shown to be an invalid laboratory sample by chain of custody measurements. From Betancourt, Dubost, Mullins, *et al.* (2007). © 2007 Society of Petroleum Engineers.

The asphaltene content of Tahiti crude oils scales well with the oil color measured by DFA (Fig. 12). As noted by Myrt E. (Bo) Cribbs of Chevron (pers. comm., 2007), the oil color measured by DFA increases with depth (Fig. 13) (Betancourt, Dubost, Mullins, *et al.*, 2007). This critical observation led to a detailed understanding of the asphaltene distribution and a basic discovery in petroleum science. The M21A (central) and M21B sands, which are not connected, show the same variation of coloration, thus asphaltene. In addition, the coloration of the M21A North sand also shows the same type of variation, except that the color is reduced substantially, indicating that the North section of M21A is not connected, a finding first identified by DFA and now acknowledged by reworking seismic and geochemistry data. Thus, DFA is addressing the biggest risk factor in this and virtually all deepwater fields, compartmentalization. The first production well (PS in Fig. 13) found the "right" color oil, indicating connectivity of the penetrated sand with the bulk of the reservoir, which is great news for the Tahiti development.

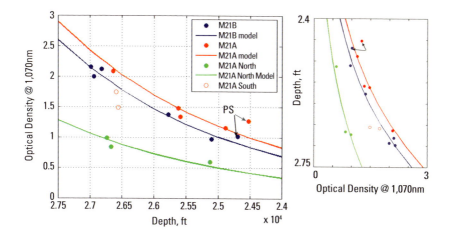

Figure 13. The Tahiti field asphaltenes are graded (their concentration depends on height in the column) even though there is little variation in GOR. The M21A and M21B sands exhibit similar color and very similar grading. However, the M21A North sand shows a very different color but with similar grading. Considering the success of the equilibrium model to account for this color grading, it appears that the M21A North section is not connected. From Betancourt, Dubost, Mullins, *et al.* (2007). © 2007 Society of Petroleum Engineers.

Color trending is not properly accounted for by the Høier-Whitson or any other EOS model from first principles. There are several reasons why:

- It has only recently been established what the likely state of asphaltene aggregation is in crude oil.
- The theory for asphaltene phase behavior is only now being widely accepted. In particular, the lack of a role for resins in this theory is now accepted.
- Most asphaltene theorists have been working on asphaltene phase behavior, with little attention paid to the asphaltene distribution in reservoirs.
- Nobody had the relevant data needed to parameterize the theoretical models.

Recent work has established that asphaltenes are nanocolloidal in toluene. The high-Q ultrasonic curve for asphaltenes in toluene is shown on the left side of Fig. 14. Aggregates form at ~150 mg/L, and the constant ultrasonic slope at concentrations greater than 150 mg/L indicates there is a single aggregate size (that is, one compressibility) (Andreatta, Bostrom, and Mullins, 2005, 2007). With prior knowledge of the forces involved (Buenrostro-Gonzalez, Groenzin, Lira-Galeana, et al., 2001), nanoaggregates are implied. The intermolecular attractive and repulsive forces are all short range. Moreover, the attractive forces are in the interior of the asphaltene molecules—in the single aromatic core per molecule—and the repulsive steric forces are dominated by the peripheral alkane substituents. Upon nanoaggregate formation, all attractive molecular real estate is encased by repulsive alkane substituents, blocking further nanoaggregate growth.

The asphaltene diffusion constants were determined by using nuclear magnetic resonance (NMR) (Freed, Lisitza, Sen, et al., 2007). At the critical nanoaggregate concentration (CNAC) the diffusion constant changes by a factor of 2, the translational diffusion constant is linear in radius, and the 2× change of the diffusion constant upon nanoaggregate formation indicates that there are ~8 molecules per nanoaggregate (Freed, Lisitza, Sen, et al., 2007). At higher concentrations, the diffusion constant remains invariant, meaning that asphaltene nanoaggregate size is invariant, which is the same conclusion from the high-Q ultrasonic measurements.

There is also an abrupt change in the hydrogen index (HI) at the same concentration. HI depends on the hydrogen content of the sample, thus its name. It also depends on the rate of relaxation of the proton spins. Below the CNAC the asphaltenes exhibit a large HI, whereas above the CNAC the asphaltenes register a reduced HI. The abrupt change in HI at the CNAC occurs because the alkane chains undergo restricted diffusion upon nanoaggregate formation, which increases spin-spin relaxation. Both the NMR translational diffusion measurements and measured transverse relaxation time (T_2) effect on the HI provide independent corroboration that the CNAC is at ~150–200 mg/L. Moreover, the restricted diffusion of the alkanes at the CNAC is expected upon nanoaggregate formation. The small change in diffusion constants proves that these are nanoaggregates.

The results from high-Q ultrasonics and NMR have been confirmed by AC conductivity experiments, with a sudden change in solution conductivity at the CNAC (Sheu, Long, and Hamza, 2007), and by centrifugation studies (Mostowfi, Indo, Mullins, et al., in press). The centrifugation study obtained an asphaltene nanoaggregate diameter of ~2.5 nm, which is very close to the ~1.6 nm obtained in the Tahiti field study, as subsequently discussed.

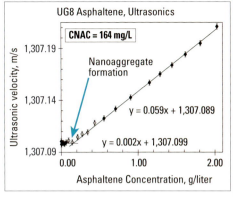
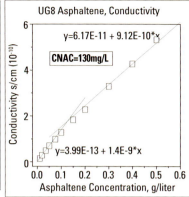

Figure 14. The ultrasonic data (left) shows the concentration of the formation of asphaltene nanoaggregates in toluene and implies a single size (Andreatta, Bostrom, and Mullins, 2005, 2007). Ionic conduction (right) of asphaltenes in toluene is used to observe the CNAC. Upon nanoaggregate formation, the conduction decreases owing to the increased Stokes drag of the ions. © 2005 American Chemical Society.

New studies on DC conductivity of asphaltenes in toluene match very closely the high-Q ultrasonics data. In addition, the aggregate number was determined to be roughly 6 in these studies (Zeng, Song, Johnson, et al., in press). The right plot in Fig. 14 shows the ionic conduction of UG8 asphaltene in toluene. The fraction of ionic components to neutrals in toluene is ~10^{-5}. The conduction at 40 Hz where the measurements are made is known to be ionic because it is in phase and frequency independent. The ionic conduction is dependent on the Stokes drag coefficient ($6\pi r\eta$) of the ion in solution, which in turn depends linearly on the radius r and viscosity η. Using a simple model transitioning from a monomer to nanoaggregate at the CNAC, the slope change gives an aggregate number of ~6 when averaging results from two different asphaltenes. Both the CNAC and the small aggregation number are consistent with other studies (Zeng, Song, Johnson, et al., in press).

With this understanding of asphaltene nanoaggregates, the asphaltene gradient was analyzed with the assumption that asphaltenes are colloidally suspended in crude oil as they are in toluene. With a solid (asphaltenes) of colloidal size suspended in a homogeneous liquid (same GOR), the following equation applies (Mullins, Betancourt, Cribbs, et al., 2007; Betancourt, Dubost, Mullins, et al., 2007):

$$\frac{OD\,(h)}{OD\,(0)} = \exp\left(-\frac{V\Delta\rho g h}{kT}\right), \tag{1}$$

where $OD(0)$ is the optical density (at some wavelength) at reference height 0, $OD(h)$ is the optical density at height h above 0, V is the volume of the asphaltene colloidal particle, $\Delta\rho$ is the density difference between the bulk liquid phase and the asphaltene particle, g is the Earth's gravitational acceleration, k is Boltzmann's constant, and T is (absolute) temperature. The term $V\Delta\rho g$ is just Archimedes buoyancy.[3] Equation 1 is simply the Boltzmann distribution for which the energy of excitation is Archimedes buoyancy multiplied by height. The Boltzmann distribution is the foundation of statistical mechanics and has enjoyed a century of confirmation.

The solid curves in Fig. 13 are from Eq. 1. Height is plotted on the horizontal axis because h is the independent variable. The insert added to Fig. 13 shows the plot rotated with height (or depth) as vertical, as is the norm in the oil field. The only adjustable variable is V. From the Tahiti data, the corresponding asphaltene colloidal diameter is ~1.6 nm, which is nanocolloidal. The remarkable finding established by this study is that asphaltenes are nanocolloidal in crude oil as they are in toluene (Mullins, Betancourt, Cribbs, *et al.*, 2007; Betancourt, Dubost, Mullins, *et al.*, 2007). The close match between the Tahiti field results and state-of-the-art petroleum science provides powerful confirmation of the methodology. Moreover, Eq. 1 can be used to predict connectivity: if the oil column has the correct functional relation of color *and* GOR, it is probably connected. An important consideration is that the governing physics controlling the GOR distribution is fundamentally different from that controlling the asphaltene distribution. All fluid components must be understood to predict reservoir connectivity, not just a single fluid property such as GOR. Each fluid component or group of components has a story to tell. That different components may be subject to different physics provides multiple stringent tests for reservoir description.

In the Tahiti case, DFA fluid color predictions were made by Betancourt, Dubost, Mullins, *et al.* (2007). The first development well accessed fluids with the color predicted by DFA, thereby indicating connectivity; Fig. 15 shows the predicted and measured log data, particularly the crude oil color. This process is the vision of exploiting DFA to the fullest. By this method, DFA provides a means to test the reservoir description. When the predictions match, the reservoir and fluid models are reinforced. When the predictions do not match, the MDT tool is still in the well, and additional measurements can be made to sort out why the expectations failed. This is far preferable to figuring out *after* the job that a problem exists. By then it is too late to make MDT measurements to uncover the source of the error. Real-time analysis—especially with predicted log data—is key.

---

[3] Archimedes, c. 287 BCE–c. 212 BCE, was living in Syracuse when King Hiero of Syracuse suspected that his goldsmith might have diluted the gold he used to make a crown for the king with other metals, keeping some gold for himself. But the crown had the correct weight for solid gold. The king asked Archimedes to determine if the crown was pure gold. While getting into his bath, Archimedes realized "Archimedes buoyancy": that the weight of an object depends upon what it is immersed in, and by this means density can be established. The oft-repeated story is that Archimedes jumped up shouting "Eureka!" ("I have found it" in Greek) and ran through the streets of Syracuse stark naked, still shouting "Eureka." It was proved that the goldsmith had stolen from the king—a mistake one makes exactly once in life.

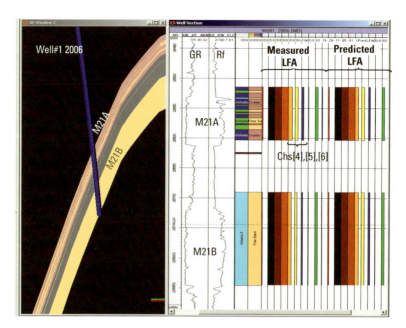

Figure 15. By knowing the reservoir geologic and fluid model, the DFA log data can be predicted and compared with log data in real time. Reasonable agreement in the comparison of DFA log predictions with actual log data implies connectivity in this case. Thus, no change in the MDT job program is indicated. From Betancourt, Dubost, Mullins, *et al.* (2007). © 2007 Society of Petroleum Engineers.

One linchpin of our DFA approach and exemplified by this asphaltene study is that it is insufficient only to establish a correlation in the field data. It is essential that the governing physics be understood and that the principles presumed to govern reservoir fluids *must* be consistent with laboratory petroleum science and reservoir understanding.

The question immediately arises as to how generally applicable these results are. Are asphaltenes always nanocolloidal, with a ~1.6-nm diameter, in all black oils? As previously discussed, current laboratory petroleum science would indicate that this is the case. A new field study has been performed on a tilted sheet reservoir of very high permeability. This study also finds asphaltene nanoaggregates of ~1.6-nm size (Betancourt, Ventura, Pomerantz, *et al.*, in press). Moreover, if the crude oil (and the corresponding asphaltenes) is in equilibrium throughout the column, then the crude oil components should have the same chemical identity throughout the column. This analysis was performed using ultrahigh-resolution mass spectroscopy and two-dimensional gas chromatography. The results are clear, that the asphaltene and crude oil chemical identities are the same, as mandated by a single column in equilibrium.

Ultrafiltration studies of heavy oil also find nanoaggregates, but some are of larger size (Zhao and Shaw, 2007), presumably because of the higher asphaltene concentration, which matches what is seen in toluene (Yudin and Anisimov, 2007). That is, a ~2-nm nanoaggregate may be applicable to all black oils, but heavier oils possibly require a modified description. This is an area of active research.

With this knowledge about asphaltene distributions in black oils where the liquid column is essentially invariant—only the colloidal concentration of the asphaltenes changes—Schlumberger is proceeding with developing a theory for the coloration of condensates in an equilibrium distribution. In condensates Høier-Whitson shows that the liquid phase exhibits variable composition with height in the column. Consequently, the changing solvation power of the liquid phase for asphaltenes most likely dominates over the gravitation term. With luck, there will be a theoretical formalism accounting for all DFA parameters.

Nonequilibrium distributions of hydrocarbons

In many reservoirs the hydrocarbon fluids are fairly accurately represented by an equilibrium model.[4] However, many reservoir fluids are very far from equilibrium. If convection is very efficient, driven either (in part) by gravity or thermal gradients, then gradients can be smaller than expected. However, frequently nonequilibrium gradients are greatly in excess of those predicted by equilibrium models. There are many reasons why these nonequilibrium conditions occur.

Transients across geologic time

Processes operative on reservoir fluids are effectively transients that are time dependent on geologic time. In the shorter time (human life) perspective, these could be called steady states.

Thermal gradients are associated with a heat flux and generate entropy. Reservoirs that have significant thermal gradients are not in equilibrium—these processes are covered by irreversible thermodynamics, whereas equilibrium implies reversibility. Generally, but not necessarily, thermal gradients result in an increased concentration of the lighter hydrocarbons in the hot regions. Large thermal gradients, such as those in geothermally active areas such as Japan, can reportedly yield large compositional gradients (Ghorayeb, Firoozabadi, and Anraku, 2003). Many reservoir models can handle thermal gradients.

---

[4] *Equilibrium is defined as no change occurs with time* and *processes are reversible.*

Biodegradation can yield large gradients. The microbes responsible for biodegradation live in water and consume alkanes, preferentially n-alkanes, at the OWC. Diffusion attempts to mix the column and remove any gradient. If the biodegradation process is active or sufficiently recent, it often overwhelms diffusive mixing. However, the microbes typically die at about 80 degC, and they become much less active as this temperature is approached. Almost all reservoirs that are colder than 80 degC have active biodegradation unless the reservoir was heated and sterilized and then uplifted and cooled. Figure 16 shows an example of biodegradation (Larter, Huang, Adams, et al., 2006). Biodegradation of crude oil is basically anaerobic and can be associated with significant methane production (Jones, Head, Gray, et al., 2008).

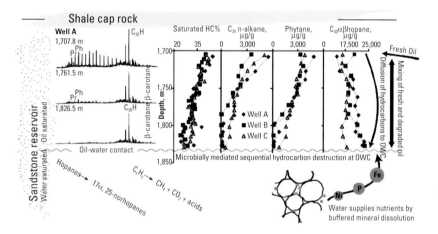

Figure 16. Saturated hydrocarbon contents and gas chromatograms of petroleum extracted from reservoir cores show a progressive increase in biodegradation in three wells from a Chinese oil field. Pr marks the pristane peak, Ph is phytane, and $C_{30}H$ is C_{30} hopane. Hydrocarbons diffuse toward the OWC, where they are degraded by microorganisms living in and using nutrients derived from the water-saturated zone below the oil column. Key nutrients, such as phosphorus, are probably buffered by mineral dissolution reactions. Fresh oil is charged to the top of the reservoir at the same time that degradation occurs. Compositional gradients reflect this complex charge and degradation scenario with bioreactive compounds (for example, normal and isoprenoid alkanes, shown as n-alkane and phytane, respectively) decreasing in concentration toward the OWC. Bioresistant compounds, such as hopanes, increase in concentration. Degradation produces new compounds, such as acids, methane, and $17\alpha,25$-norhopanes, which get distributed between the oil and water phases. From Larter, Huang, Adams, et al. (2006). © AAPG 2006, reprinted by permission of the AAPG whose permission is required for further use.

Figure 16 shows that the concentration of all saturated alkanes in general and the C_{25} n-alkane concentration in particular reduce to very low values toward the OWC. The reduction of alkane content toward the bottom of the oil column results in an increase of asphaltene content. This causes a huge variation of viscosity in the column from top to bottom, with an order of magnitude not uncommon. The phytane content is similarly reduced. Phytane, a diterpenoid alkane (or tetra isoprenoid), 2,6,10,14-tetramethylhexadecane to be exact, can also be consumed by microbes. In contrast, the hopane biomarker shown in Fig. 16 is resistant to biodegradation and remains in high and slightly increasing concentration toward the OWC (Larter, Huang, Adams, et al., 2006). The gas chromatography traces in Fig. 16 show the classic signature of biodegradation, the loss of n-alkanes (which appear as the spikes in the top trace). Figure 16 shows all these trends are in three wells, implying fieldwide variations (Larter, Huang, Adams, et al., 2006). In spite of the importance of biodegradation in this example, there is still need for invoking charge history (fresh oil charge in Fig. 16) as a component to understanding this reservoir (Larter, Huang, Adams, et al., 2006). Charge history is treated in greater detail later in this chapter. The extent of biodegradation is often a big factor in determining oil quality and can vary with distance to the OWC, fault block, depth, or sand and within individual reservoir sands within a single structural compartment. The MDT sample acquisition and analysis program for DFA can play a major role in revealing the complexities induced by biodegradation.

The variation of viscosity with biodegradation occurs in large measure because of the corresponding variation of asphaltene content. The variation of viscosity with asphaltene content is legend. In the example shown in Fig. 17, the variation of viscosity with asphaltene content is 6 orders of magnitude, and yes, this work was conducted on asphalt paving materials in addition to maltenes (Lin, Lumsford, Glover, et al., 1998). There is little difference in Fig. 17 between the viscosity of maltene plus asphaltene versus the viscosity of asphalt. Maltene is defined as deasphaltened crude oil, so it relates directly to dead reservoir crude oils. A modified Pal-Rhodes viscosity model based on the Einstein viscosity equation (with solvation) is shown to fit the data for the zero-shear-rate, complex, dynamic viscosity reasonably well. Similar large variations of the viscosity of asphaltene-rich systems occur with temperature variations (Sirota and Lin, 2007).

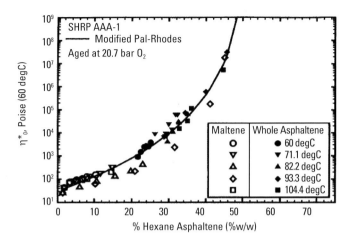

Figure 17. The viscosities of both asphalt and maltene depend enormously on the asphaltene content (maltene is dead, deasphaltened crude oil). From Lin, Lumsford, Glover, *et al.* (1998). With kind permission of Springer Science+Business Media.

Current reservoir charging: Hydrocarbons

Another mechanism that produces huge gradients is the real-time charging of reservoirs. Again, in general, the reservoir fluid dynamic process only needs to be faster than diffusive mixing to create large nonequilibrium conditions in the reservoir. Such processes are clocked in geologic time. An excellent example uncovered by Elshahawi, Hows, Dong, *et al.* (2007) is a massive influx of biogenic methane into a reservoir that contained a black oil (Fig. 18). As noted by Dan McKinney of Shell (pers. comm., 2007), the variation in the isotopic ratio of $^{13}C/^{12}C$ immediately shows that the reservoir is not in equilibrium and that the increase in biogenic methane is associated with a large increase in GOR (Elshahawi, Hows, Dong, *et al.*, 2007).

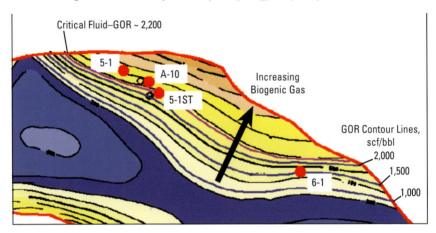

Figure 18. Large variations in GOR (as shown by the GOR contour lines) are induced by a massive influx of biogenic methane into an oil reservoir. The reservoir fluids are not in equilibrium, as shown by the variable ratio of the stable carbon isotopes (compare with Fig. 19). From Elshahawi, Hows, Dong, *et al.* (2007). © 2007 Society of Petroleum Engineers.

Biogenic and thermogenic methane are distinguished by carbon isotope ratios and to some degree by the gas composition. Carbon has two stable isotopes, ^{12}C and ^{13}C in natural abundances of 99% and 1%, respectively. In contrast ^{14}C is radioactive and is used in carbon dating.[5] The isotope ratio of a sample is defined in Eq. 2 and is reported in units of per mil (‰):

$$\delta^{13}C_{sample} = \left(\frac{\left[\frac{^{13}C}{^{12}C}\right]_{sample}}{\left[\frac{^{13}C}{^{12}C}\right]_{standard}} - 1 \right) \times 1{,}000. \qquad (2)$$

The common standard reference for $\delta^{13}C$, the Chicago PDB Marine Carbonate Standard (CPDB), was obtained from the Cretaceous marine fossil *Belemnitella americana*, from the PeeDee formation in South Carolina, which has a generally accepted absolute ratio of $^{13}C/^{12}C$ of 0.0112372. This material has a higher $^{13}C/^{12}C$ ratio than nearly all other natural carbon-based substances; for convenience, it is assigned a $\delta^{13}C$ value of zero, which gives almost all other naturally occurring samples negative delta values (J. Zumberge, pers. comm., 2008).

Plants preferentially take up CO_2 containing the lighter carbon isotope. In large accumulations, microbially produced (biogenic) gases consist almost exclusively of methane (Jones, Head, Gray, et al., 2008) and have $C_1/(C_2 + C_3)$ concentration ratios greater than 1,000 and $\delta^{13}C$ CPDB values of methane more negative than –60 ‰. Petroleum-related (thermogenic) hydrocarbon gases resulting from kerogen catagenesis generally have $C_1/(C_2 + C_3)$ ratios smaller than 50 and $\delta^{13}C$ CPDB values of methane more positive than –50 ‰ (J. Zumberge, pers. comm., 2008).

The $\delta^{13}C$ for the field depicted in Fig. 18 is shown in Fig. 19. Biogenic methane clearly increases up structure in the field; thus, the reservoir fluids are necessarily not in equilibrium.

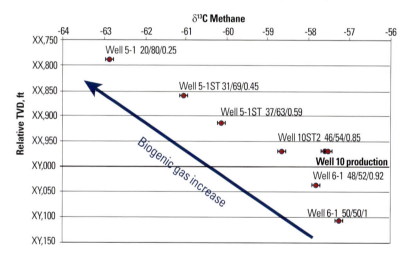

Figure 19. Biogenic methane monotonically increases updip at the field shown in Fig. 18. The values by each well are the % thermogenic gas, % biogenic gas, and their ratio, respectively. The variable isotope ratios indicate that the reservoir fluids are not in thermodynamic equilibrium. From Elshahawi, Hows, Dong, et al. (2007). © 2007 Society of Petroleum Engineers.

[5] Neutron flux in the upper atmosphere causes the nuclear reaction $^{14}N + n \rightarrow {}^{14}C + p$. Any bioactivity that consumes carbon consequently ingests ^{14}C, which decays with a half-life of ~5,730 years. Upon death, consumption stops, so measuring the $^{14}C/^{12}C$ ratio dates the time since death. W.F. Libby invented "carbon dating" within a week of hearing about the neutron flux from cosmic rays during a presentation on upper atmosphere physics and was awarded the Nobel Prize in Chemistry in 1960. (One of the author's former advisors, Professor T. Harrison Davies, was sitting next to Libby at this talk when Libby first pondered the fate of this large flux of neutrons by leaning over and asking, "I wonder what becomes of all those neutrons." Unfortunately, this inquiry did not pique the curiosity of my advisor.)

Although the primary mechanism for fluid compositional variation in Figs. 18 and 19 is probably a large biogenic charge, there is also a strong variation of the dead black oil. Figure 20 shows that not only does the GOR vary, which is most likely due primarily to the secondary biogenic methane charge, but the asphaltene content of the live oil and the API gravity of the dead oil also show large variations. The dead oil variation may be due to variable kinetics of solvation of differing oil fractions into the light ends up structure. Dead crude oil variations can be associated with asphaltene destabilization and flocculation associated with large methane injection (Fujisawa, Mullins, Dong, et al., 2003). However, in this case, Fig. 20 shows that there is a large variation of API gravity for dead crude oils obtained from live crude oils of the same GOR. Thus, asphaltene destabilization from differential methane admixing cannot explain the dead crude oil variations. Additional mechanisms producing dead crude oil variations associated with charge history and described in the following may also be in play (Stainforth, 2004). Interpolating and extrapolating oil properties away from wells, a key feature in understanding production, mandates this understanding. DFA is a key technology to help distinguish the relative importance of different reservoir processes (Mullins, Elshahawi, and Stainforth, 2008).

It is important to comprehend that the oil component with by far the largest diffusion constant, methane, is often not in equilibrium across a field. Even more profound is that a nonequilibrium distribution of methane can also exist within a single sand within a single well (Elshahawi, Venkataramanan, McKinney, et al., 2006; Mullins, Elshahawi, Hashem, et al., 2005). In addition to showing compartmentalization, which is addressed later (see "Compartments"), Fig. 21 shows that the $\delta^{13}C$ ratio is not constant in the single, clean sand depicted. The reservoir is being charged by biogenic methane; it is plausible that the biogenic methane tends to override the oil as it migrates updip. This explanation is basically consistent with the small but measurable GOR and $\delta^{13}C$ gradients. The analysis showing a changing pressure gradient is consistent with the measured compositional profile. Such a consistency check will be markedly improved with DFA density measurements, as previously discussed in "DFA density measurements."

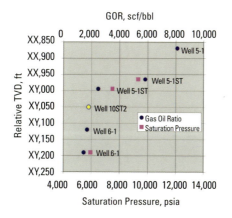
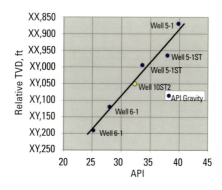
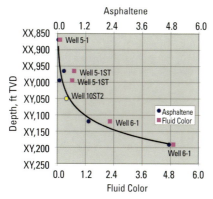

Figure 20. The fluid properties of the reservoir depicted in Fig. 18 show a large variation in GOR and saturation pressure resulting from a secondary biogenic methane charge. The variation in API gravity of the dead oil and of the asphaltene content evidently result from another mechanism. From Elshahawi, Hows, Dong, et al. (2007). © 2007 Society of Petroleum Engineers.

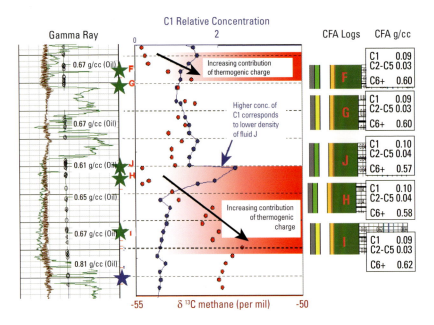

Figure 21. CFA measurements (on the right) from five DFA stations show that the fluid at station G is higher density than that at station J; there is a density inversion, with higher density fluids higher in the column; thus the intervening shale is probably sealing. The gentle compositional gradient from stations J to I is confirmed by a decrease in methane concentration in the mud-gas as well as an isotope trend that indicates an increasing thermogenic contribution with depth. Gamma ray and formation pressure data are in the left panel. The fluids from DFA stations J, H, and I are not in equilibrium as gleaned from the isotope data. After Dong, Elshahawi, Mullins, *et al.* (2007) and Mullins, Elshahawi, Hashem, *et al.* (2005). © 2005 and © 2007 Society of Petroleum Engineers.

Current reservoir charging: Nonhydrocarbons

CO_2, H_2S, and N_2 can all be present in significant quantities naturally as well as injected for pressure maintenance and increasingly for sequestration. DFA methods can directly detect CO_2 and indirectly detect N_2 in many circumstances. H_2S is not within the DFA scope at present, but work is ongoing. All these nonhydrocarbon gases can charge reservoirs in very different processes than hydrocarbon generation and charging. The CO_2 source location, migratory path, and charge history all can differ from those of hydrocarbons.

For example, CO_2 can be sourced from magma and often charges reservoirs near plate boundaries. (The $\delta^{13}C$ ratios are useful for understanding the origins of CO_2, with the $\delta^{13}C$ for magmatic CO_2 tending to be negative by several ‰.) DFA has been used to detect CO_2 in reservoirs near the plate boundary depicted in Fig. 22 (Müller, Elshahawi, Dong, *et al.*, 2006), with reservoir CO_2 concentrations detected up to 22% in corresponding West African fields. CO_2 is also of great concern in Southeast Asian fields as well.

These reservoirs are presumably currently charging in the same manner that Lake Nyos in Cameroon is currently charging. A plume of CO_2 erupted from Lake Nyos in 1986, killing almost 1,800 people. CO_2 enters at depth in Lake Nyos; the hydrostatic head pressure and lack of water strata mixing enable the deep waters to become enriched in CO_2. (The water column, similar to many oil columns, is grossly out of equilibrium.) Any disturbance can cause the lake waters to overturn and release a giant cloud of CO_2. Because CO_2 is heavier than air (N_2 and O_2), the CO_2 cloud hugs the ground in a blanketing descent down valleys and suffocates the life out of any being it encounters. Lake Nyos is currently charging with CO_2 and is being continuously vented to prevent future deadly eruptions, as shown in the lower inset on Fig. 22. A simple gas lift process is used.

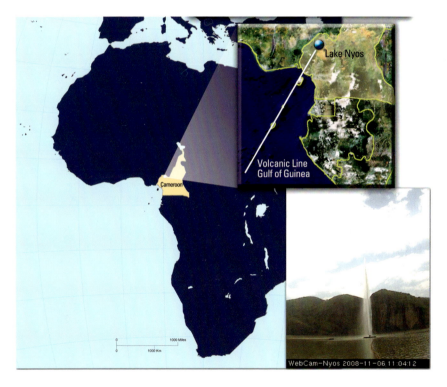

Figure 22. Magmatic CO_2 is currently charging into oil reservoirs and lakes near this plate boundary (white line) in West Africa. From Müller, Elshahawi, Dong, et al. (2006). © 2006 Society of Petroleum Engineers. The inset shows the current (February 5, 2008) venting of CO_2 out of Lake Nyos by a gas lift process. Photograph courtesy of http://pagesperso-orange.fr/mhalb/nyos/nyos.htm.

Some reservoirs are ~100% CO_2, and N_2 can also be present. In a remarkable observation, K.C. Khong of Schlumberger (pers. comm., 2007) notes that the ternary diagram of fluids from a reservoir near the volcanic Hainan Island in China shows that the CH_4:N_2 ratio is fixed whereas the CO_2 concentration is highly variable (Fig. 23) (Xu, Cai, Guo, *et al.*, 2008). The CO_2:N_2 ratio could be expected to be fixed if they are both magmatic in origin. It is not understood at this time (at least by this author) why N_2 and CH_4 scale.

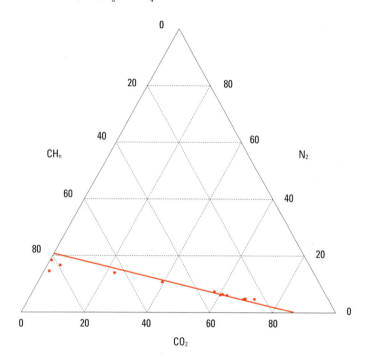

Figure 23. The gas composition is plotted in a ternary diagram for reservoirs in a geologically active basin. The CH_n:N_2 ratio is nearly constant, a surprising and unexplained result. The CO_2 concentration varies widely and is presumably magmatic in origin. From Xu, Cai, Guo, *et al.* (2008).

Reservoir charge history

Reservoir hydrocarbons exhibit a broad array of fluid types in large measure because of the diversity of kerogen types: oil prone (Type I), oil and gas prone (Type II), and gas prone (Type III) (Killops and Killops, 2005; Tissot and Welte, 1984). Kerogen is the insoluble, organic component of rocks, a solid that is insoluble in any solvent. Within this context, bitumen is often defined as the soluble organic component of source rock (as opposed to reservoir rock) and is present in source rocks in much smaller quantities than kerogen. Heat-induced chemical reactions are required to liberate fluid hydrocarbons from the solid kerogen. The "oil window" of kerogen is variously listed ranging from 60 degC to 120 degC whereas the "gas window" ranges from 120 degC to 150 degC (Killops and Killops, 2005; Tissot and Welte, 1984). In general, kerogen is sourced from marine or lacustrine (occasionally terrestrial) organic matter that quickly settled into anoxic basins. Kerogen types are conveniently plotted on the van Krevelen diagram in terms of the atomic ratios of H/C versus O/C (Fig. 24). In addition, identifiable plant-derived constituents or macerals are used to characterize kerogens. Macerals are the organic equivalent to minerals for rock composition.

Type I kerogen is less common and exhibits a very high atomic H/C ratio and a low O/C ratio (Killops and Killops, 2005; Tissot and Welte, 1984). Often rich in alginite from freshwater algae, Type I kerogen contains substantial lipid fractions and low aromatic fractions. This type has the highest oil-producing potential compared with other kerogen types (Killops and Killops, 2005; Tissot and Welte, 1984). The well-known Green River kerogen in the United States is an example of an immature Type I lacustrine kerogen that still holds its massive oil potential. Type II, rich in the macerals exinite, cutinite, resinite, and liptinite, typically originates in marine sediments and is by far the most common type of kerogen. In addition to its aliphatic moieties, it contains more aromatic carbon (thus less hydrogen) and somewhat more oxygen on average than Type I (Killops and Killops, 2005; Tissot and Welte, 1984). Sulfur is often present in significant amounts in Type II kerogens, and high-sulfur Type II kerogens are often separated into the subtype Type II-S when the S/C ratio exceeds 0.04. Type III kerogen, rich in vitrinite, is often sourced from woody terrestrial organic matter, particularly from vascular plants lacking in lipids or waxy matter (Tissot and Welte, 1984; Killops and Killops, 2005). Consequently, Type III kerogen is gas prone. In particular, terrestrial environments generally have much more exposure to oxygen than marine environments; the corresponding oxidation of the organic matter removes hydrogen and increases the aromaticity of the carbon. In addition, terrestrial plants, unlike marine plants, contain a rigid cell wall in part constructed from lignin, a robust and highly aromatic chemical component. That is, terrestrial organic matter initially contains much more aromatic carbon than marine organic matter, and terrestrial organic matter is more readily oxidized, further increasing the aromatic carbon content. High-rank coals represent an extreme of highly aromatic carbon with little hydrogen, whereas oil contains mostly alkane carbon. Highly aromatic carbonaceous materials are solid, such as coal and graphite. The most aromatic component of oil, asphaltene, is a friable, infusible solid with roughly half aromatic carbon, half alkane carbon. Thus, in a sense, asphaltenes are intermediate between (the liquid fraction of) crude oil and coal. Finally, methane, such as coalbed methane, is commonly associated with terrestrial organic matter. Type IV kerogen is rich in inertinite and has no ability to generate oil and gas. This kerogen is thought to originate from deposition of charred organic matter.

In addition to carbon, hydrogen, and oxygen of concern in Fig. 24, sulfur is another key constituent. The extent of sulfur aromaticity has been shown to track the extent of carbon aromaticity for a wide variety of carbonaceous resources (Mitra-Kirtley and Mullins, 2007). Various biological and thermal processes can give rise to H_2S. For kerogen in shales, any H_2S produced tends to get scavenged by iron in the shales. For kerogen in carbonates, iron is lacking and the H_2S often ends up with the hydrocarbons in the reservoir. Different kerogens, their chemical environments, and their evolution produce the huge array of hydrocarbons and other components in the subsurface.

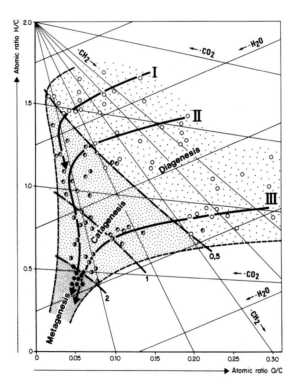

Figure 24. Different kerogen types, their chemical evolution during maturation, and vitrinite reflectance values (0.5, 1, 2) are shown on the van Krevelen diagram. Type I kerogen is oil prone, Type II is oil and gas prone, and Type III is gas prone. From Tissot and Welte (1984). With kind permission of Springer Science+Business Media.

For a given kerogen, the heating rate (generally the rate of burial) and cooking time (duration at particular temperatures) strongly affect the type of produced hydrocarbon. Figure 25 is a typical plot of the types of hydrocarbons produced in kerogen catagenesis. The figure refers to the Kimmeridge clay kerogen, which is responsible for North Sea oil and gas. Kerogen catagenesis is a disproportionation reaction producing two reaction products, the hydrocarbon fluid and a mature kerogen. Kerogen enters the oil-generation window at ~60 degC. The lowest temperatures and least cracking produce the heaviest hydrocarbon. Athabasca bitumen, with an API gravity of ~8, is so shallow it is mined. Microbes typically biodegrade shallow oils, thereby decreasing the API gravity; nevertheless, the heaviest hydrocarbon generation is for kerogen just entering the oil window. As the kerogen is heated at higher temperatures for longer times, cracking continues and lighter hydrocarbons are produced. This is similar to refinery operations; extensive heating at high temperatures produces more cracking and thus lighter hydrocarbons (along with coke). The gas window for kerogen is at ~120 degC. Therefore, if a kerogen is generating hydrocarbons at variable temperatures during burial, then variable hydrocarbons are charged into the reservoir. In a normal burial sequence, lighter hydrocarbons are charged at later times into the reservoir.

For both kerogen and cracking in refinery operations, the overall chemical reaction is a disproportionation reaction. That is, the single source material creates two very different reaction products: one is very heavy and solid whereas the other is a fluid. The kerogen case starts with immature kerogen and ends with a mature kerogen and a fluid hydrocarbon. The mature kerogen is much more aromatic than immature kerogen, having both lost alkane carbon and been carbonized. The maturity of a kerogen is measured by its vitrinite reflectance; the more mature kerogen has more aromatic vitrinite, which is darker in color and correspondingly more reflective. In the refinery case, a hydrocarbon feedstock is separated by distillation into various weight liquid fractions. Cracking in refining (essentially pyrolitic reactions) yields coke, an insoluble black carbonaceous material, along with some fluid components. If cracking occurs with reservoir hydrocarbons, the end result is a solid pyrobitumen and a lighter hydrocarbon than the original fluid. Extensive cracking takes place at high temperatures, yielding methane. Methane can be either thermogenic, and typically deep, or biogenic, and typically shallow. Often reservoirs have contributions to methane from both sources. As discussed previously, the methane isotope ratio is used to differentiate the source of methane.

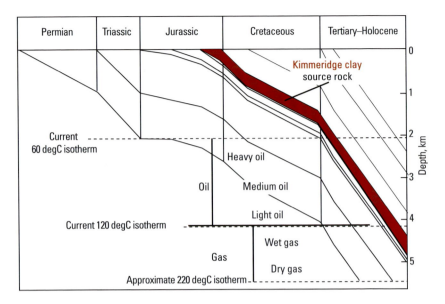

Figure 25. Different hydrocarbons are produced from a kerogen (here Kimmeridge clay) as a function of depth, thus time and temperature. Greater temperatures for longer times result in more cracking, producing lighter hydrocarbons.

There has been an assumption in basin modeling that reservoir fluids are in equilibrium, possibly coupled with subtle effects from a thermal gradient. In the reservoir engineering community, this assumption was made in part because many reservoir models cannot handle any greater complexity. Moreover, prior to DFA there was no cost-effective means to elucidate potential fluid complexity in the reservoir, especially at early times in reservoir evaluation. Without direct proof to the contrary, some technologists tended to view the limitations of the models as limits on natural fluid variability—the simple models were taken literally. Moreover, in basin modeling, it has been viewed as inviolable that reservoirs are charged with variable fluids and then equilibrated. This conceptual model does acknowledge that variable fluids arise in the kerogens. The assertion that equilibrium prevails is founded on the fluids being in the reservoirs for geologic time. If it is presumed that there are no current dynamic effects on the reservoir fluids, then geologic time is operative to establish equilibrium. Moreover, in a normal burial sequence, lighter hydrocarbons enter the trap at later times (Fig. 25). If the hydrocarbon charge enters at depth in the reservoir, then the lighter hydrocarbons would migrate through heavier hydrocarbons as they rise to the top, thereby enhancing mixing and thus equilibrium.

However, recent work examining fluids in many reservoirs suggests that frequently the memory of the charge history is retained by the reservoir oil (Stainforth, 2004). When this occurs, the oil column can be far out of equilibrium and the gradients are generally much *larger* than expected in equilibrium models. Figure 26 shows many reservoirs that are not in equilibrium and mandate a more involved explanation. As noted by Stainforth (2004) for the depth versus GOR plotted for various reservoirs, the compositional variations are not necessarily "more or less linear with depth unless the reservoirs are near critical," as predicted by the equilibrium conditions of a gravitationally induced gradient of Høier and Whitson (2001). Consequently, other mechanisms inducing gradients are operative. Figure 26 shows that variations apply both to the GOR of the live oils as well as the API gravity of the dead oils.

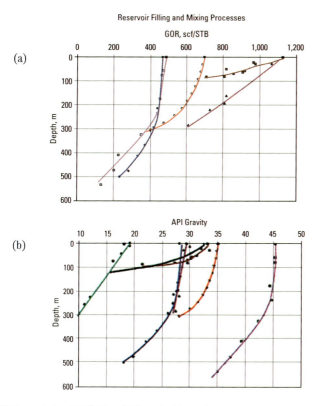

Figure 26. The variation of GOR (a) and API gravity (b) are shown versus depth for several reservoirs. Even though these reservoirs have low-GOR fluids, they also exhibit large compositional variations that curve with depth. That is, for equilibrium conditions, these relatively incompressible fluids should exhibit small GOR gradients. Consequently, nonequilibrium is indicated. From Stainforth (2004). Used with permission of The Geological Society of London.

The schematic of the model used to describe this and other data is shown in Fig. 27 (Stainforth, 2004). The concepts are a bit involved. The overriding assumption is that the fluids do not mix well in the trap; there is no magic stirring bar that mixes the fluids. In many cases, fluids enter a trap in a very local, spatially isolated way. There is not a broad fluid front that enters the trap from the bottom. If there were, mixing would be enhanced. But for fluids entering a high-conductivity streak, such as a fracture, mixing is impaired. Within the Stainforth model, the fluids enter the trap in successive layers, with later-charging light hydrocarbons displacing the heavier charge in place, in the manner of piston displacement.

In the Stainforth (2004) model depicted schematically in Fig. 27, the heavy ends are controlled by their sequential yield of heavy ends from the kerogen. Thus, the color variation reflects the asphaltene content from the kerogen. The GOR can be controlled by a corresponding charge mechanism if the gas does not break out as a separate phase. That is, the GOR can be controlled by the sequential gas yield from the kerogen. However, if there is a gas cap, no matter how small, the GOR of the hydrocarbons can be controlled in a much different process. The hydrocarbons in the reservoir are presumed to be in equilibrium with the gas cap at the GOC only. Liquids below the GOC are not in equilibrium with the gas cap (in this model). The extent of the dissolved gas is controlled mostly by the temperature and pressure at that GOC at that point in time. For a normal burial sequence, as new liquids enter the trap, the liquids previously at the GOC are pushed down and that GOR is preserved. That is, the new liquids, being low density, rise to the top, thus pushing existing liquids down. As burial continues, the (gas cap) pressures increase, thereby increasing the GOR for the fluids that charge later in the life of a trap. There are variants to this model, but the foundation is given here. The light ends and the heavy ends can be under fundamentally different controls. Moreover, equilibrium could be reached with the light ends but not the asphaltenes. For example, the diffusion constant of methane is much larger than the diffusion constant of asphaltene nanoaggregates in crude oil. The full suite of DFA measurements must be brought to bear to assess the state of reservoir hydrocarbons. This dictates a new approach to designing MDT jobs by employing basin modeling to understand the possible fluid variations and conducting DFA analysis to find which of the family of possible fluid curves is correct (Mullins, Elshahawi, and Stainforth, 2008).

Often many physics mechanisms are at play in determining the fluid distribution. For example, large heavy-end gradients could be due to both the Stainforth charge history mechanism as well as biodegradation. It is possible that previously identified gradients were incorrectly identified as being due to biodegradation and could actually be Stainforth columns (J. Stainforth, pers. comm., 2007). How could this misidentification be made?

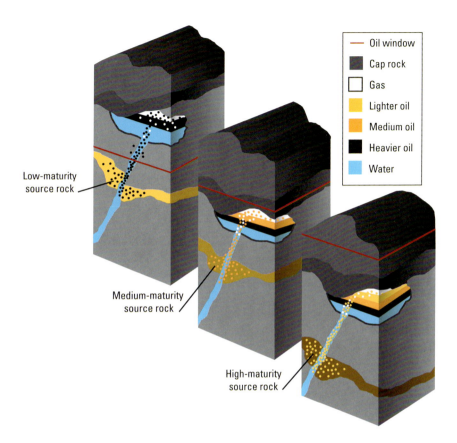

Figure 27. Charge history determines the hydrocarbon distribution in the Stainforth (2004) model. Heavy ends are controlled by their temporal production from the kerogen whereas light ends can be controlled by historical PVT conditions at successive GOCs. The resulting fluids are often grossly out of equilibrium.

Figure 28 shows the generation of biomarkers during kerogen catagenesis (Wilhelms and Larter, 2004; Tissot and Welte, 1984). The generation of biomarkers occurs primarily with the initial charge and varies in oils by 5 orders of magnitude in reservoir hydrocarbons (Wilhelms and Larter, 2004). In the case of multiple reservoir charging, the biomarkers can be dominated by the heaviest charge, whereas the fluid properties relevant to production are dependent on all the charges. Consequently, because it is now understood that different components in the oil can be under the control of different mechanisms, and the very small biomarker fraction may not reflect the dominating physics for the bulk of the oil. The importance of understanding the dominant physics controlling the distribution of reservoir fluids cannot be overstated. Often the reconnaissance data of the reservoir is very limited, especially before production. Consequently, extensive interpolation and extrapolation of fluid and other parameters is routine. The resolution of the relative importance of biodegradation and charge history must be determined for individual reservoirs; DFA provides the basis for the requisite data acquisition to resolve these fundamental issues.

In any event, Fig. 28 convincingly argues that DFA is a much better way to characterize the bulk of the oil than biomarker analysis. That is, comparing light ends to heavy ends, such as by measuring GOR and asphaltene content (with color measurements), compares the bulk properties of the oil and is much less likely to be affected by potentially misleading analyses of components present in the small mass fraction.

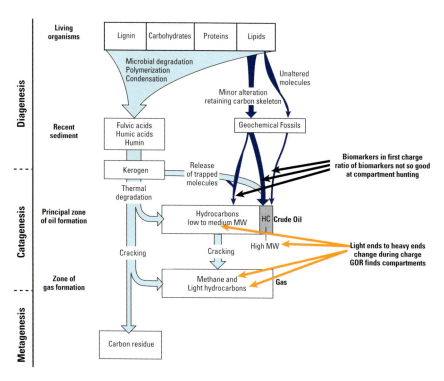

Figure 28. Various crude oil components are generated at different maturation stages. The bulk of the biomarkers are produced early on in the kerogen catagenesis process. Compartments are best found by comparing ratios of components produced early *and* late in the charge history. After Wilhelms and Larter (2004). Used with permission of The Geological Society of London.

Of course, other factors can induce compositional variations, including water washing and leaky seals. Over time these and other mechanisms will be brought into the fore with DFA as the vehicle to investigate existing compositional variations in the reservoir in a cost-effective manner. The growing body of DFA case studies elucidates the varied and sundry physics controls on reservoir fluid variations and portends an increasing understanding of reservoir fluids, a fascinating topic of critical importance.

Compartments

Compartmentalization is the biggest problem in deepwater today for virtually all operators. The problem can be summarized in some tongue-in-cheek "advice" that Laurence P. Dake gives to young people in his book *The Practice of Reservoir Engineering* (2001) in reference to offshore projects:

> *Sooner or later (usually later) the project phase comes to an end . . ., all the equipment has been hooked-up and the development phase commences. In terms of career advice, this is as good a time as any for the reservoir engineer to transfer from one company to another, preferably a long way away, for now the truth is about to be revealed.*

The origin of this run-and-hide advice is that for offshore projects, there is no opportunity to do history matching before big resources are put at risk. The entire project must be forward modeled and is based on little data because of the expense of penetrating offshore fields with wells and the huge costs of well testing. For example, the recently announced Jack field had only one well test. Without history matching or even extended well testing, compartment volumes are not measured and can be *much* smaller than anticipated. Moreover, proper interpretation in history matching and especially well testing is an art; to make accurate predictions from these interpretations is challenging.

This author notes with all due respect to the enormously productive L.P. Dake that the quoted statement is horrendous advice to give to young people aspiring to greatness. We can do much better by using new technology such as DFA to address reservoir complexities and reduce risk. The health of the oil business or any other discipline of human endeavor relies critically on attracting skilled and motivated young people to build on the foundations established by their predecessors.

Moreover, onshore and offshore reservoirs are fundamentally similar. Onshore reservoirs have not posed the same degree of extreme risk as offshore because it is possible to perform some history matching for the former during their more modest capital expenditure phase. Nevertheless, with an increasing focus on efficiency and reduction of time to large production, it is no longer appropriate to launch any projects by employing horribly flawed geologic and fluid models with the presumption of subsequent corrections. Proper planning of the technological and economic implications of projects increasingly requires accurate reservoir understanding.

Compartmentalization looms as the source of enormous, costly errors in the petroleum industry because of three primary problems.

Compartments: Problem 1

There is no physics known to humankind that can possibly image at reservoir-length scale, detect sealing barriers invisible to wireline petrophysical measurements, and perform this task at a distance, through miles of earth.

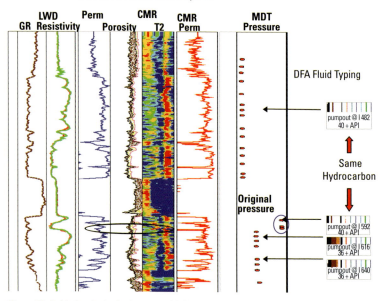

Figure 29. A thin barrier in the lower sand is identified by DFA, showing a stairstep discontinuous change of fluid properties at the barrier. The barrier is not indicated from petrophysical wireline logs although it is holding off 2,000 psi of depletion pressure drop (M. Hashem, pers. comm., 2004). From Mullins, Fujisawa, Elshahawi, *et al.* (2005). © 2005 Society of Petroleum Engineers.

Figure 29 shows an example of a sealing barrier that is holding off 2,000 psi of depletion pressure drop and yet is invisible to wireline petrophysical logging (M. Hashem, pers. comm., 2004; Mullins, Fujisawa, Elshahawi, *et al.*, 2005). The fluids change discontinuously at the barrier, so DFA methods detect the barrier easily, provided that a sufficient number of DFA stations are performed.

Compartments: Problem 2

The industry has settled into the practice of presuming that pressure communication means flow communication or "connectivity." Thus, if the pressure gradients line up in two wells, the corresponding permeable zones in the two wells are presumed to be in flow communication. It almost sounds reasonable: "pressure communication means flow communication." However, this presumption is very inaccurate and can be off by 9 orders of magnitude—a factor of 1 billion. Pressure communication is a necessary but *insufficient* condition for establishing flow communication. Pressure communication can occur on a geologic timescale whereas flow communication must occur on a production timescale. This is a contrast of 10 million years with 10 years, respectively, a difference of 6 orders of magnitude. Moreover, pressure communication can occur through very low permeability, a conceptual pinhole, whereas flow communication requires permeability. This is another few orders of magnitude difference. Consequently, the standard industry practice for establishing flow communication can be in error by a factor of 1 billion. The proof that this solution of compartment identification is horribly flawed is that compartmentalization remains one of the biggest problems in the oil industry.

Compartments: Problem 3

The industry takes it for granted that compartments should be big. Only if there is proof that compartments are small is this then acknowledged. However, the initial presumption that compartments are big violates all geostatistical measurements ever performed on the frequency of events versus the magnitude of events. For example, the 100-year storm is *bigger* than the 10-year storm, not smaller. Figure 30 shows a typical geostatistical measurement, this time on earthquakes. Bigger earthquakes are much less frequent than smaller earthquakes, as seen by the linear slope of –1 on the log-log plot. This relationship is seen to apply over many decades, with roll-off for the small-magnitude earthquakes because the really small ones are too difficult to measure and, mercifully, from the lack of data at the large-magnitude end. The expectation that compartments are big would have a frequency versus magnitude plot of +1 slope (or at least a positive slope), which is exactly the opposite of all geostatistical measurements.

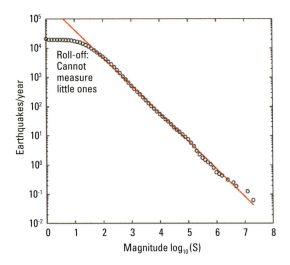

Figure 30. The frequency versus magnitude of earthquakes on a log-log plot produces a straight line with a –1 slope over many decades. *All* geophysical objects show similar scaling for the relationship between the frequency of events and their magnitude. Correspondingly, small compartments should be much more frequent than large compartments. Assumptions that compartments are big violate geostatistics.

An example of a small compartment discovered by Mohamed Hashem of Shell is shown in Fig. 31. When this author quizzes people on how many barrels of oil could be in place in this noneconomic sand, deepwater Gulf of Mexico, generally the answer comes back in the 1 million to 10 million bbl range. The expectation is that compartments are big. However, this compartment is ~600 bbl (Elshahawi, Hashem, Mullins, *et al.*, 2005; Mullins. Fujisawa, Elshahawi, *et al.*, 2005). Indeed, the first reaction of many in the industry upon hearing about a 600-bbl compartment is "well, that can't be right, the data must be misinterpreted," with the implicit thought that compartments must be big.

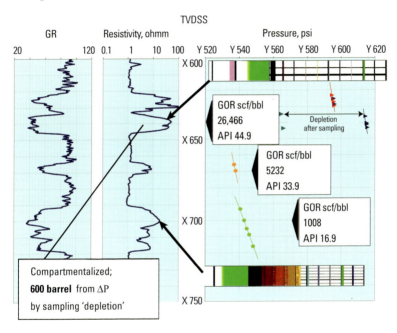

Figure 31. This 600-bbl compartment was depleted by 50 psi after ~½ bbl was removed during sampling. The pressure gauge was checked; the measurement is correct. Small compartments should be expected; for any given oil-bearing zone, large compartment volumes need to be proved, not assumed. From Elshahawi, Hashem, Mullins, *et al.* (2005) and Mullins, Fujisawa, Elshahawi, *et al.* (2005). © 2005 Society of Petroleum Engineers.

After "producing" a ½ bbl during sampling, the pressure in the sand was reduced by 50 psi. The pressure gauge was checked first by going to the sand zone above the sampled zone (two purple points in the upper right area of the pressure plot in Fig. 31). The same pressures and gradient before and after sampling were obtained. A second point was measured in the depleted zone, and the same 50-psi pressure drop was recorded. In other words, the pressure gradient remained the same but the pressure dropped as a result of removal of the ½ bbl of fluid. By estimating fluid compressibility and with a known sampling volume and pressure drop, the hydrocarbon in place is estimated to be 600 bbl.

Some have expressed surprise that such a small compartment could have been intersected by a well at 20,000 ft. This question continues to exemplify the misunderstanding of the standard geophysical concepts in Fig. 30. There isn't just one 600-bbl compartment—there are lots of them. Small compartments are numerous; it is the big ones that are rare. The 600-bbl compartment was intersected by a well because they are so numerous that "you couldn't miss."

A corollary to the geophysical frequency versus magnitude plot of Fig. 30 is that geophysical objects typically scale in dimension. For example, an unusually large turbidity current is most likely thick, laterally extensive, and longitudinally extensive. However, it is standard practice to consider the height of a sand to be a "layer" thickness, with no implication of the layer thickness on the layer lateral dimensions. If DFA proves that a compartment height is one-half of what it was thought to be, then it is plausible that both the length and width of the compartment are also halved in comparison with previous expectation. Thus, the volume of the compartment is divided by 8—essentially an order of magnitude. Figure 32 shows an example of a geophysical object scaling, with the thickness and width of river and nearshore clastics roughly scaling over several decades (Reynolds, 1999).

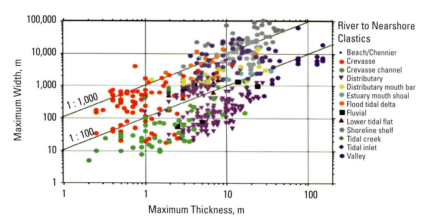

Figure 32. The river and nearshore clastics have a scaling relationship. The thicker deposits tend to be more laterally extensive. Geophysical objects often scale; dividing the height by 2 (for example, as proved by DFA measurement) can divide the expected volume by 8. From Reynolds (1999). © AAPG 1999, reprinted by permission of the AAPG whose permission is required for further use.

The geophysical relations of both frequency versus magnitude and scaling of physical dimensions also apply to faults (Elshahawi, Hashem, Mullins, *et al.*, 2005). There are many reservoirs that are naturally fractured; these geophysical concepts clearly apply directly to oil reservoirs. The bottom plot in Fig. 33 shows that bigger faults are less frequent, with the familiar linear scaling on a log-log plot. The top plot of Fig. 33 shows a definite relationship between fault length and throw—the width of the distribution of the length/throw ratio is very small, again establishing the scaling of spatial dimensions of geophysical objects. Fault length is roughly 16 times the fault throw.

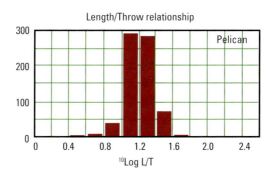

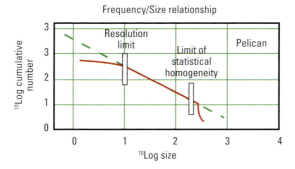

Figure 33. Geophysical scaling relationships apply to faults. Top: Fault throw scales with fault length, and the corresponding width of the distribution is small. Bottom: Larger faults are much less frequent. From Elshahawi, Hashem, Mullins, *et al.* (2005). © 2005 Society of Petroleum Engineers.

These three factors—incommensurate physics, pressure communication not necessarily meaning flow communication, and industry expectations for large compartments that grossly violate geostatistical observations—combine to make compartmentalization the biggest problem for virtually every operator offshore. Deepwater reveals deficiencies in current practices of reservoir exploitation. The same deficiencies apply to land production but they remain more easily hidden in land production. Nevertheless, there is no fundamental distinction in reservoir physics for land versus offshore. Efficient production of oil and gas, an increasingly important objective, mandates that reservoir physics be understood and treated properly in reservoir reconnaissance.

Another prevailing trend for land, shelf, and deepwater is that the target reservoirs are often at increasing depths; the shallower and easier crude oil has already been exploited. Of course, the overall pressures, both overburden and pore pressures, increase with depth. Compaction naturally occurs with increasing overburden and gives rise to two pernicious conditions, decreased porosity and increased sealing. Conductivity paths can be squashed, creating lower or no connectivity between different reservoir units. Consequently, the gargantuan difficulties that the oil industry has experienced with compartments are expected to increase significantly in the future as deeper reservoirs are produced. In addition, there is no question that deeper reservoirs especially in deeper water are associated with a much higher cost structure than their shallower cousins. With the increasing costs in this environment, there will be significant pressure to reduce the extent of reservoir evaluation. This coupling of much poorer reservoir quality with much higher cost structure means that standard formation evaluation methods of the past are now simply antiquated and inadequate. It is more imperative than ever that new methods such as DFA coupled with new workflows be exploited to mitigate the growing risk.

DFA can readily identify compartmentalization in new ways. Higher density fluids that are higher in an oil column clearly identify likely vertical compartmentalization or stacked reservoirs (Fig. 21). As described by Dong, Elshahawi, Mullins, *et al.* (2007), Fig. 34 shows a giant density inversion, with a much higher GOR in the lower sand than in the upper sand. This clear delineation of compartmentalization is also known from the pressure data. Asphaltene content can likewise identify compartments, with much more asphaltene content and thus coloration higher in the column. Asphaltenes are the most dense components in crude oil; if anything they sink, not float (Mullins, Rodgers, Weinheber, *et al.*, 2007). Statistical methods can be used to assess whether the color and GOR differences are significant (Venkataramanan, Weinheber, Mullins, *et al.*, 2006).

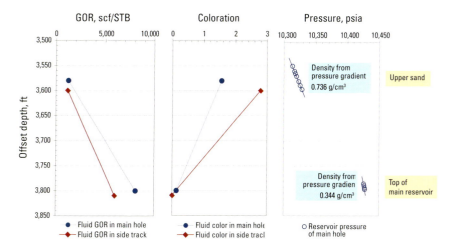

Figure 34. Fluid density inversions and color inversions determined with DFA can be used to identify compartmentalization. Here, pressure measurements and DFA prove the lack of connectivity. From Dong, Elshahawi, Mullins, *et al.* (2007). © 2007 Society of Petroleum Engineers.

The question arises as to how these fluid density inversions occur in the reservoir. It is not just random. As previously noted for a normal burial sequence, lower density hydrocarbons come out of the kerogen at later times. The range of variation of these hydrocarbons can be enormous, going from extraheavy oil at very shallow depths to deep, dry thermogenic gas (Fig. 25).

Extraheavy oil has a density of 1 g/cm^3, whereas deep, dry gas has a density on the order of 0.2 g/cm^3; this is a fluid density inversion. Indeed, measuring density provides a range of a factor of 5 over many kilometers of depth to detect fluid density inversion. However, this range is not so big per unit of vertical offset. The measurement of GOR has a much bigger range. GOR is a good proxy for fluid density, especially for oils within a given field: lower density hydrocarbons generally have a higher GOR. The range of GOR is from nearly zero for extraheavy oils to infinity for dry gas. This is the biggest range mathematically allowable (if not worrying about the *different* orders of infinity noted by Georg Cantor[6]). With the infinite GOR range over kilometers for detecting fluid density inversion, density inversions on a meter-length scale are measurable by GOR. In addition, even subtle asphaltene content inversions, more color higher in the column, are

---

[6] *Mathematician Georg Cantor, 1845–1918, showed that not all infinities are equal. He developed simple proofs showing that real numbers are a countable infinity whereas irrational numbers are not countable and thus a higher order infinity. He expended himself unsuccessfully trying to prove whether the two orders of infinities are adjacent. This issue, still unresolved, is thought to have contributed to his suicide.*

readily measured because optical color is a robust measurement. In particular, for GOR or color measured by DFA in a single well, the measurements have the same time, same temperature, same calibration, same tool, same field engineer, but different GOR or color. The measured difference in GOR between two DFA stations is rather precise, even if the absolute measurement of GOR by DFA has a significant offset. Laboratory measurements of GOR have absolute errors for each bottle (sample transfer problems, bottle leakage, etc.), so differences in GOR measured in the laboratory for different bottles are not so precise.

OBM contamination impedes comparative data analysis from different DFA stations. Nevertheless, OBM filtrates tend to lack dissolved gas and color. Consequently, increasing OBM filtrate contamination in a single crude oil reduces color and reduces GOR concomitantly. In contrast, for real variations of reservoir fluids, reduced GOR is associated with increased color.

Kerogens enter the oil window at a temperature of about 60 degC, which corresponds to relatively shallow depth. Expelled hydrocarbons generally float up, meaning that the reservoir is even colder and shallower. The key point is that often the reservoir is not fully formed as it is being charged with hydrocarbons. This scenario is not like a completely constructed house being filled with furniture; instead, the house is under construction as it is being filled. Figure 35 shows the implications of this dynamic rock *and* fluid process.

Other mechanisms can give rise to fluid density inversion. Figure 36 shows a lower trap charge spilling over into a higher trap. In this case a density inversion would occur and indicate separate compartments.

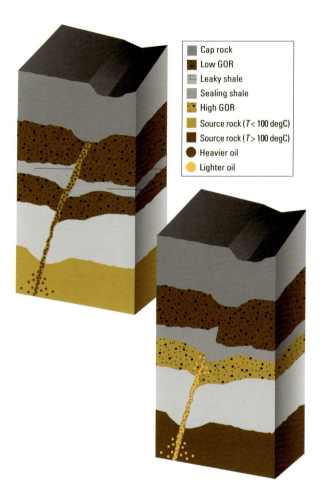

Figure 35. The oil window of kerogen is in the range ~60 degC to 120 degC, so the source rock and, more importantly, the reservoir are relatively shallow and at low pressure. In this scenario the first oil fills sand bodies below and above a leaky shale. With greater burial, the shale compacts and becomes sealing. Subsequent lighter hydrocarbons become trapped in the lower sand only, creating a fluid density inversion.

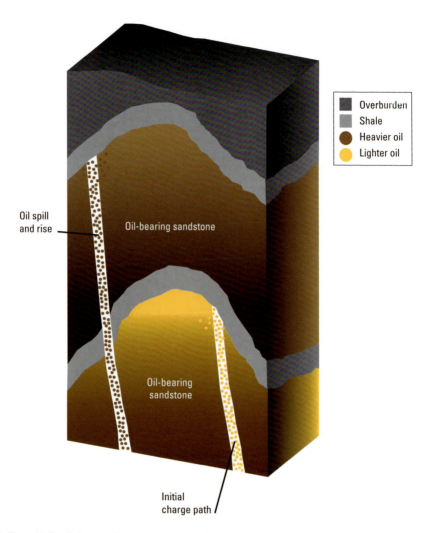

Figure 36. As oil charges a lower trap, heavier oil (color coded) spills out of the trap from the base of the column and charges a higher trap. A fluid density inversion results.

Production and miscible injection

Crude oils are delicate. Anything that happens to a crude oil can cause it to undergo a phase transition. For example, dropping the pressure on a crude oil or condensate can result in gas evolution, retrograde dew evolution, asphaltene onset, or multiple transitions. Injection of separator gas or even just methane can destabilize asphaltenes. Moreover, both pressure reduction and miscible injection can lead to large compositional variations, even if none existed before production.

The first CFA job showed four black oil zones being subjected to gas injection (Fujisawa, Mullins, Dong, et al., 2003). A monitor well was drilled to find the gas; the gas had reached the monitor well location in three of the four zones. One of the three zones had been cleanly swept at the monitor well while the other two zones were only partially swept (Fujisawa, Mullins, Dong, et al., 2003). However, the CFA measurements also identified that as the amount of dissolved separator gas increased, the asphaltene fraction decreased. For the thoroughly swept zone, the mobile fluid was colorless, indicating that this fluid contained no asphaltenes. The partially swept zones contained some color (asphaltenes) but less than the original black oil in the fourth zone. It appears likely that the separator gas had caused a phase transition yielding a mobile and an immobile phase. Of course, the asphaltenes are largely responsible for reducing mobility. Heavy oil can have coexisting in equilibrium four hydrocarbon phases: one solid, two immiscible liquids, and a gas (Shaw and Zou, 2007). Monitoring reservoir fluids for unintended consequences is important within a production vantage. There are opportunities for DFA in development for understanding sweep efficiency and phase behavior associated with miscible flooding and production.

Flow assurance and compositional variation: Asphaltenes

Most of the chemical focus for flow assurance has been directed at laboratory studies of phase behavior. Phase behavior versus temperature, pressure, and the nature of the fluids has frequently been determined for oilfield samples, frequently MDT samples. Mitigation by various treatment chemicals is determined sometimes in an Edisonian approach of trial and error. Recommendations are then made to the operating company.

A missing element in this routine approach is understanding the implications on flow assurance of fluid compositional variation (potentially in part resulting from compartmentalization). Often, the presumption is made that little fluid variation exists in obtaining a sample of "the oil" for detailed laboratory analysis. Indeed, detailed analyses are needed, for example, as described for asphaltenes (Hammami and Ratulowski, 2007). An overview of field and laboratory considerations for asphaltenes has been described elsewhere (Akbarzadeh, Hammami, Kharrat, et al., 2007). Here, we address a common occurrence—an asphaltene plugging problem—with recognition at the outset of compositional variation along with production complexities. Although there are ongoing projects with these considerations, there is nothing yet in the open literature.

So, a generic situation is described here. Generalization of this approach to any flow assurance problem is not difficult to develop.

Asphaltene deposition is a global problem affecting production in specific fields. In the field with an asphaltene problem shown in Fig. 37, the plugging is updip from water injection, downdip from gas injection, and in an area above a tar mat. The cause of the plugging could be due to one or more factors such as commingling, compositional variations, mobilization of a tar mat (many are actually powdery), depressurization, separator gas injection, and water injection complexity. DFA is required to figure out the origin of the problem. For example, if tar mat mobilization is the problem, then suspended tar mat or asphaltene particulates can easily be seen as light scattering (Hammami, Phelps, Monger-McClure, et al., 2000; Joshi, Mullins, Jamaluddin, et al., 2001) and would intensify near the base of the column only in the area of the field with the tar mat. Problems induced by commingling would imply significant fluid heterogeneity, which likewise can be revealed by DFA. Problems associated with separator gas injection would link extrahigh GOR in the wellbore or formation with asphaltene phase instability, which has been reported in DFA case studies (Fujisawa, Mullins, Dong, et al., 2003). Of course, depressurization problems would show up in laboratory reports and potentially in DFA log data. Moreover, contrasting MDT logs from "good" and "bad" areas of the field is required to validate the proposed origins of the problem. It is essential that the identified origin of the problem be present only in the part of the field exhibiting the asphaltene problem. If the same fluid conditions apply to areas exhibiting no asphaltene problem, then the origin of the problem is not correctly identified. Any conclusions regarding the origin should be consistent for DFA and laboratory data. A coordinated field study of DFA and MDT sampling coupled with advanced laboratory studies is optimal to prevent or mitigate unexpected and costly flow assurance problems.

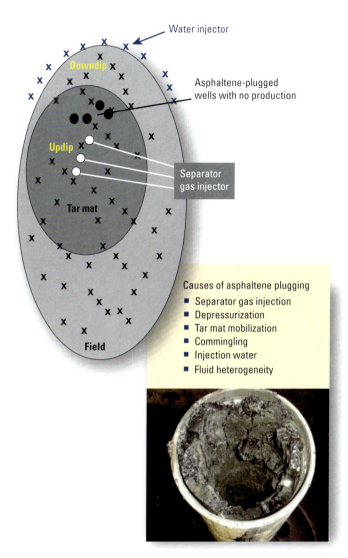

Figure 37. A field has problems with asphaltene plugging. The origin of the problem is unknown and requires DFA to resolve. The solution would then be built on this knowledge and monitored in part with DFA.

pH and water chemistry

In many fields there are substantial transition zones where both oil and water are movable. Transition zones can have large heights of 50 m or more, especially in carbonates, so understanding what will flow when producing from transition zones is important. Changing wettability—thus changing critical saturation and relative permeabilities—coupled with the difficulty of determining wettability mandates flow measurements versus height in the transition zone to determine what will flow. In such a setting, it is critical to differentiate between water-base mud (WBM) filtrate and connate water. Because both are often highly conductive, corresponding measurements are of limited utility for differentiating the waters.

Raghuraman, O'Keefe, Eriksen, *et al.* (2005) and Raghuraman, Xian, Carnegie, *et al.* (2005) showed that pH can be measured downhole by measuring the color of litmus paper dyes injected into the MDT flowline. The downhole pH measurements are more accurate than laboratory pH measurements and truly reflect downhole waters. Laboratory pH measurements are performed at 1 atm and room temperature. Reducing the pressure on the sample liberates gas. Any CO_2 or even H_2S that escapes alters the pH. Opening a can of soda exhibits this pH-altering degassing phenomenon. In addition, pH-sensitive solids such as any carbonate can precipitate out of solution with reduced temperature, thereby altering pH. Consequently, laboratory measurements are of very limited utility. Moreover, the accuracy of downhole pH measurements can be further improved through temperature, pressure, and ionic strength corrections (Raghuraman, Gustavson, Mullins, *et al.*, 2006).

In methods pioneered by Andrew Carnegie and ChengGang Xian of Schlumberger (Raghuraman, Xian, Carnegie, *et al.*, 2005), pH measurements can be used to differentiate WBM filtrate and connate water, thus enabling measurement of the oil-water cut versus height in the transition zone. Figure 38 depicts locating the OWC in a well with a substantial transition zone (Raghuraman, Xian, Carnegie, *et al.*, 2005). In favorable cases, injected water and connate water can be differentiated to provide important feedback on sweep efficiency and optimization.

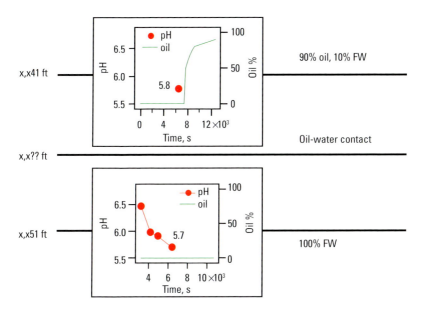

Figure 38. The difference in pH between formation water (FW) and water-base mud (WBM) filtrate can aid in the characterization of oil/water transition zones. At a depth of X,X51 ft, DFA shows 100% water and the DFA pH data shows a cleanup trend with the pH approaching that of true formation water. At the station in the transition zone, 10 ft above, DFA initially shows 100% water until about 7,443 s, when oil breaks through. The oil fraction increases to 90% by the end of sampling. However, volume fraction analysis alone cannot distinguish if this water is formation water or WBM filtrate. The pH measurement at 6,452 s indicates that the water is a mixture of formation water and filtrate, indicating that the formation water is mobile at that depth. This puts the true OWC between these two depths. From Raghuraman, Xian, Carnegie, *et al.* (2005). © 2005 Society of Petroleum Engineers.

DFA workflows

A typical deepwater workflow is to make a discovery, perform delineation drilling, stop drilling, build a geologic model, port the geologic model to a simple reservoir (dynamic) model, fill the reservoir model with a simple fluid model, perform many simulations of the reservoir on a computer using the simplistic models of the reservoir and the contained fluids, build a development plan, then construct seafloor facilities and commence development drilling. As noted by the quote from L.P. Dake (2001) in the preceding "Compartments" section, there are frequent large errors between expectations versus actual production. This workflow is unscientific.[7]

We do not all need to be scientists, but it was the scientific method that put men on the moon; engineering projects must use the scientific method. The typical workflow suffers from three major flaws:

- The reservoir model is not used to predict anything until it is too late. Testing the reservoir model *after* development drilling means that costly facilities and wells are already in place *before* predictions can be tested. The scientific method mandates not only explaining previously acquired data but predicting new observations to be tested. Just as it is easy to "predict" yesterday's stock market, it is altogether much different to predict the stock market of tomorrow.

- Incredibly, the most accurate description of the reservoir—the most accurate fluids model populating the geologic model—is generally not constructed. Billions of dollars are put at risk without ever building the most accurate model of the reservoir. This is an obvious fatal flaw in current practices and needs to be rectified immediately. The origin of this problem is plausibly the focus of the asset team on draining the reservoir. Consequently, the dynamic model is presumed to hold the key. In fact, the static model is really the key. The foundation of the dynamic model is the static model. And the static model can be constructed with all known complexity while the dynamic model is generally simplified for acceptable iteration times. It is the static model that should be used for predictive purposes and tested certainly with DFA.

---

[7] *An example of the scientific workflow is given by perhaps the greatest scientist ever, Albert Einstein. When he devised the general theory of relativity, he predicted "yesterday's data": the precession of the perihelion of the planet Mercury. This observation had been known yet unexplained for a long time. According to Einstein, the prescribed ellipse of Mercury should precess around the sun because of the slight (relativistic) increase in mass as Mercury nears the sun, which increases the gravitational attraction. Most importantly, Einstein also predicted that starlight would bend when passing near massive objects. This bold prediction was confirmed by Sir Arthur Stanley Eddington's expedition to measure the bending of light during a solar eclipse in 1919. The success of this bold prediction made Einstein a household name.*

- Moreover, the typical deepwater workflow determines the error in predictions by computer simulations, not by measurement. The reservoir is a physical object; errors in understanding reservoir structure *must* be measured. To exacerbate this problem, the simulations are time-consuming and dependent on the complexity incorporated from both the geologic model and the fluids model. In order to run many simulated production expectations and for error prediction, the reservoir models tend to incorporate simplified fluid models and simplified geologic models, which are often grossly in error, further reducing the effectiveness of the already flawed approach of determining error on a computer. This author cannot think of any other experimental discipline where error is determined solely on a computer. In experimental disciplines, error is measured. And, as discussed previously, with increasing target depth, there is an increasing cost structure, increasing reservoir complexity, and decreasing opportunity for reservoir evaluation (fewer wells). New workflows are mandated.

Well testing to test the reservoir model is sometimes performed. Nevertheless, the cost of a well test is often very large especially in a deepwater setting. Consequently, for given reservoirs, well testing is performed infrequently, and in deepwater settings the well tests are often terminated early because of mounting expenses. For understanding reservoirs well testing is valuable but does not nearly represent a complete solution.

The new DFA workflow

The recommended workflow begins with building the most accurate description of the reservoir (at that current time). Use this model to predict DFA log data. Perform DFA measurements to test the accuracies of the predictions. Most importantly, the comparisons between predicted and acquired logs must be performed in real time while the MDT tool is in the well. If predictions are in agreement with measurements, then the current model is supported and the MDT tool can be pulled out of the hole. If the DFA log predictions differ from acquired log data, the MDT tool is in the well and measurements can be made as to the source of the error. In some cases, models can be updated in real time and can thus generate new predictions that can be tested concurrently by the MDT tool. For example, EOS fluid models can be updated in real time. Subsequent to this process, the geologic and fluid models can be updated and new wireline tests devised to test the updated model.

This entire predictive workflow has been performed on the Tahiti field, with the log data matching predictions (Fig. 15). The color analysis of the crude oil confirmed that the first development well is placed in sands that are probably connected with the major reservoirs of the field (Betancourt, Dubost, Mullins, *et al.*, 2007). This workflow should be replicated on a regular basis to extract the full utility from the MDT runs and to provide much more stringent experimental testing of geologic and fluid models.

The geologic and fluid models should be built early in the process. These models can even be constructed based on basin modeling and thus precede any wireline data acquisition (Mullins, Elshahawi, and Stainforth, 2008). In this case, basin modeling can be used to develop a family of fluid curves. These fluid curves are parametric in critical parameters such as kerogen type, heating rate, and mixing of reservoir fluids. Other complexities can be incorporated such as biodegradation and multiple charges. The DFA job can then be designed to limit the possible range of these parameters, thereby providing valuable information about the reservoir history. That is, the volume of allowable parameters in parameter phase space can be reduced by the wireline job. The overall fluid properties of the field, plus the specific variations of fluid properties throughout the field, are useful for addressing basin modeling predictions. Building the models early enables optimized wireline logging and reservoir evaluation while testing the accurate static geologic and fluid models. Such an approach to an evergreen model with continuous model predictions, measurement, error determination, and updating enables the best understanding of the reservoir to be developed in the most efficient manner.

Senior technologists

This last section is short and at the end of the chapter. However, make no mistake—without the attention of the senior technologists in real time, the entire potential of great reservoir understanding through DFA is significantly curtailed. The scientific workflow advocated earlier mandates that DFA log analysis be performed in real time. During the job, it is essential that experienced senior technologists, such as Schlumberger Reservoir Domain Champions (RDCs), are responsible for making the assessment whether data acquisition is proper and whether the resulting analysis is in agreement with expectations. When the expected log data do not match predictions, the MDT tool is still in the well and additional measurements can be made to reveal the source of the disagreement. The senior technologist has the necessary expertise and experience to determine what new MDT measurements are required for DFA to resolve the discrepancy. Moreover, this process is best performed in conjunction with senior technologists in the operating company in order to incorporate the best understanding of the reservoir. This requirement deviates from some typical petrophysical logging workflows, in which a skilled wireline field engineer acquires log data from a scripted program and the resulting data is analyzed at some later time offsite. It is of paramount importance that the crucial real-time analytic role performed by the senior technologists be recognized.

Conclusions

Reservoir fluids exhibit all manner of complexities for manifold reasons. In the past, many of the factors inducing fluid complexity remained unrealized, both because the physical and chemical processes inducing complexities of reservoir fluids were not appreciated and because there was no method to reveal these complexities. The most cost-effective means to reveal the secrets of reservoirs and their contained fluids is the expanding new DFA technology. The applications of DFA are steadily growing worldwide, on a wide variety of wells, reservoirs, and fluid types. Revealing the complexities of reservoirs and their contained fluids by using DFA technology is a proven vision, and the increasing arsenal of DFA services will expand the ability to gain even more insight.

To further optimize the DFA workflow, there are increasing numbers of senior technologists both in Schlumberger and in operating companies who are capable of overseeing the MDT logging runs from the vantage point of addressing reservoir concerns by exploiting the expanding DFA technology. The spirit of this community is infectious. Young people can enter this fold and make a difference, especially with the new tools and workflows. Future prospects are bright, and we are optimistic as we address the biggest technical uncertainty in the oil industry, the reservoir.

References

Akbarzadeh, K., Hammami, A., Kharrat, A., Zhang, D., Allenson, S., Creek, J., Kabir, S., Jamaluddin, A., Marshall, A.G., Rodgers, R.P., Mullins, O.C., and Solbakke, T.: "Asphaltenes—Problematic but Rich in Potential," *Oilfield Review* (Summer 2007) 19, No. 2, 22–43.

Andreatta, G., Bostrom, N., and Mullins, O.C.: "Ultrasonic Spectroscopy of Asphaltene Aggregation," *Asphaltenes, Heavy Oils, and Petroleomics*, O.C. Mullins, E.Y. Sheu, A. Hammami, and A. Marshall (eds.), New York, New York, USA, Springer (2007), 231–258.

Andreatta, G., Bostrom, N., and Mullins, O.C.: "High-Q Ultrasonic Determination of the Critical Nanoaggregate Concentration of Asphaltenes and the Critical Micelle Concentration of Standard Surfactants," *Langmuir* (2005) 21, 2728–2736.

Betancourt, S.S., Dubost, F.X., Mullins, O.C., Cribbs, M.E., Creek, J.L., and Mathews, S.G.: "Predicting Downhole Fluid Analysis Logs to Investigate Reservoir Connectivity," paper SPE 11488 presented at the International Petroleum Technology Conference, Dubai, UAE (December 4–6, 2007).

Betancourt, S.S., Fujisawa, G., Mullins, O.C., Eriksen, K.O., Dong, C., Pop, J., and Carnegie, A.: "Exploration Applications of Downhole Measurement of Crude Oil Composition and Fluorescence," paper SPE 87011 presented at the SPE Asia Pacific Conference on Integrated Modelling for Asset Management, Kuala Lumpur, Malaysia (March 29–30, 2004).

Betancourt, S.S., Ventura, G.T., Pomerantz, A.E., Viloria, O., Dubost, F.X., Zuo, J., Monson, G., Bustamante, D., Purcell, J.M., Nelson, R.K., Rodgers, R.P., Reddy, C.M., Marshall, A.G., and Mullins, O.C.: "Nanoaggregates of Asphaltenes in a Reservoir Crude Oil," *Energy & Fuels* (in press).

Buenrostro-Gonzalez, E., Groenzin, H., Lira-Galeana, C., and Mullins, O.C.: "The Overriding Chemical Principles that Define Asphaltenes," *Energy & Fuels* (2001) 15, No. 4, 972–978.

Carnegie, A.: "Understanding the Pressure Gradients Improves Production from Oil/Water Transition Carbonate Zones," paper SPE 99240 presented at the SPE/DOE Symposium on Improved Oil Recovery, Tulsa, Oklahoma, USA (April 22–25, 2006).

Dake, L.P.: *The Practice of Reservoir Engineering* (rev. ed.), Amsterdam, The Netherlands, Elsevier Science (2001), p. 10.

Dong, C., Elshahawi, H., Mullins, O.C., Venkataramanan, L., Hows, M., McKinney, D., Flannery, M., and Hashem, M.: "Improved Interpretation of Reservoir Architecture and Fluid Contacts Through the Integration of Downhole Fluid Analysis with Geochemical and Mud Gas Analyses," paper SPE 109683 presented at the Asia Pacific Oil and Gas Conference and Exhibition, Jakarta, Indonesia (October 30–November 1, 2007).

Dubost, F.X., Carnegie, A.J., Mullins, O.C., O'Keefe, M., Betancourt, S.S., Zuo, J.Y., and Eriksen, K.O.: "Integration of In-Situ Fluid Measurements for Pressure Gradients Calculations," paper SPE 108494 presented at the International Oil Conference and Exhibition, Veracruz, Mexico (June 27–30, 2007).

Elshahawi, H., Hashem, M.N., Mullins, O.C., and Fujisawa, G.: "The Missing Link— Identification of Reservoir Compartmentalization Through Downhole Fluid Analysis," paper SPE 94709 presented at the Annual Technical Conference and Exhibition, Dallas, Texas, USA (October 9–12, 2005).

Elshahawi, H., Hows, M., Dong, C., Venkataramanan, L., Mullins, O.C., McKinney, D., Flannery, M., and Hashem, M.: "Integration of Geochemical, Mud-Gas, and Downhole-Fluid Analyses for the Assessment of Compositional Grading—Case Studies," paper SPE 109684 presented at the SPE Annual Technical Conference and Exhibition, Anaheim, California, USA (November 11–14, 2007).

Elshahawi, H., Venkataramanan, L., McKinney, D., Flannery, M., Mullins, O.C., and Hashem, M.: "Combining Continuous Fluid Typing, Wireline Formation Tester, and Geochemical Measurements for an Improved Understanding of Reservoir Architecture," paper SPE 100740 presented at the SPE Annual Technical Conference and Exhibition, San Antonio, Texas, USA (September 24–27, 2006).

England, W.A.: "The Organic Geochemistry of Petroleum Reservoirs," *Advances in Organic Geochemistry* (1990) 16, 415–425.

England, W.A., Muggeridge, A.H., Clifford, P.J., and Tang, Z.: "Modelling Density-Driven Mixing Rates in Petroleum Reservoirs on Geological Time-Scales, with Application to the Detection of Barriers in the Forties Field (UKCS)," *The Geochemistry of Reservoirs*, J.M. Cubitt and W.A. England (eds.), London, England, Geological Society of London SP 86 (1995), 185–201.

Freed, D.E., Lisitza, N.V., Sen, P.N., and Song, Y.-Q.: "Molecular Composition and Dynamics of Oils from Diffusion Measurements," *Asphaltenes, Heavy Oils, and Petroleomics*, O.C. Mullins, E.Y. Sheu, A. Hammami, and A.G. Marshall (eds.), New York, New York, USA, Springer (2007), 279–300.

Fujisawa, G., Betancourt, S.S., Mullins, O.C., Torgersen, T., O'Keefe, M., Terabayashi, T., Dong, C., and Eriksen, K.O.: "Large Hydrocarbon Compositional Gradient Revealed by In-Situ Optical Spectroscopy," paper SPE 89704 presented at the SPE Annual Technical Conference and Exhibition, Houston, Texas, USA (September 26–29, 2004).

Fujisawa, G., Mullins, O.C., Dong, C., Carnegie, A., Betancourt, S.S., Terabayashi, T., Yoshida, S., Jaramillo, A.R., and Haggag, M.: "Analyzing Reservoir Fluid Composition In-Situ in Real Time: Case Study in a Carbonate Reservoir," paper SPE 84092 presented at the SPE Annual Technical Conference and Exhibition, Denver, Colorado, USA (October 5–8, 2003).

Ghorayeb, K., Firoozabadi, A., and Anraku, T.: "Interpretation of the Unusual Fluid Distribution in the Yufutsu Gas-Condensate Field," *SPE Journal* (2003) 8, No. 2, 114–123.

Hammami, A., Phelps, C.H., Monger-McClure, T., and Little, T.M.: "Asphaltene Precipitation from Live Oils: An Experimental Investigation of Onset Conditions and Reversibility," *Energy & Fuels* (2000) 14, No. 1, 14–18.

Hammami, A., and Ratulowski, J.: "Precipitation and Deposition of Asphaltenes in Production Systems: A Flow Assurance Overview," *Asphaltenes, Heavy Oils, and Petroleomics*, O.C. Mullins, E.Y. Sheu, A. Hammami, and A.G. Marshall (eds.), New York, New York, USA, Springer (2007), 617–660.

Høier, L.: "Miscibility Variations in Compositionally Grading Petroleum Reservoirs," PhD thesis, Norwegian University of Science and Technology, Trondheim, Norway (1997).

Høier, L., and Whitson, C.H.: "Compositional Grading—Theory and Practice," *SPE Reservoir Evaluation & Engineering* (2001) 4, No. 6, 525–535; also presented as paper SPE 63085 at the SPE Annual Technical Conference and Exhibition, Dallas, Texas, USA, (October 1–4, 2001).

Jones, D.M., Head, I.M., Gray, N.D., Adams, J.J., Rowan, A.K., Aitken, C.M., Bennett, B., Huang, H., Brown, A., Bowler, B.F.J., Oldenburg, T., Erdmann, M., and Larter, S.R.: "Crude Oil Biodegradation via Methanogenesis in Subsurface Petroleum Reservoirs," *Nature* (2008) 451, 176–180.

Joshi, N.B., Mullins, O.C., Jamaluddin, A., Creek, J., and McFadden, J.: "Asphaltene Precipitation from Live Crude Oil," *Energy & Fuels* (2001) 15, No. 4, 979–986.

Killops, S.D., and Killops, V.J.: *An Introduction to Organic Geochemistry* (2nd ed.), Malden, Massachusetts, USA, Blackwell Publishing (2005).

Larter, S., Huang, H., Adams, J., Bennett, B., Jokanola, O., Oldenburg, T., Jones, M., Head, I., Riediger, C., and Fowler, M.: "The Controls on the Composition of Biodegraded Oils in the Deep Subsurface: Part II—Geological Controls on Subsurface Biodegradation Fluxes and Constraints on Reservoir-Fluid Property Prediction," *AAPG Bulletin* (June 2006) 90, No. 6, 921–938.

Lin, M.S., Lumsford, K.M., Glover, C.J., Davison, R.R., and Bullin, J.A.: "The Effects of Asphaltenes on the Chemical and Physical Characteristics of Asphalt," *Asphaltenes, Fundamentals and Applications*, E.Y. Sheu and O.C. Mullins (eds.), New York, New York, USA, Plenum Press (1998).

Mitra-Kirtley, S., and Mullins, O.C.: "Sulfur Chemical Moieties in Carbonaceous Materials," *Asphaltenes, Heavy Oils, and Petroleomics*, O.C. Mullins, E.Y. Sheu, A. Hammami, and A.G. Marshall (eds.), New York, New York, USA, Springer (2007), 157–188.

Mostowfi, F., Indo, K., Mullins, O.C., and McFarlane, R.: "Critical Nanoaggregate Concentration of Asphaltenes by Centrifugation," *Energy & Fuels* (in press).

Müller, N., Elshahawi, H., Dong, C., Mullins, O.C., Flannery, M., Ardila, M., Weinheber, P., and McDade, E.C.: "Quantification of Carbon Dioxide Using Downhole Wireline Formation Tester Measurements," paper SPE 100739, presented at the SPE Annual Technical Conference and Exhibition, San Antonio, Texas, USA (September 24–27, 2006).

Mullins, O.C., Betancourt, S.S., Cribbs, M.E., Dubost, F.X., Creek, J.L., Andrews, A.B., and Venkataramanan, L.: "The Colloidal Structure of Crude Oil and the Structure of Oil Reservoirs," *Energy & Fuels* (2007) 21, No. 5, 2785–2794.

Mullins, O.C., Elshahawi, H., Hashem, M.N., and Fujisawa, G.: "Identification of Vertical Compartmentalization and Compositional Grading by Downhole Fluid Analysis: Towards a Continuous Downhole Fluid Log," *Transactions of the SPWLA 46th Annual Logging Symposium*, New Orleans, Louisiana, USA (June 26–29, 2005), paper K.

Mullins, O.C., Elshahawi, H., and Stainforth, J.S.: "Use of Basin Modeling to Optimize Wireline Fluid Analysis," *Transactions of the SPWLA 49th Annual Logging Symposium*, Edinburgh, Scotland (May 26–28, 2008), paper NN.

Mullins, O.C., Fujisawa, G., Elshahawi, H., and Hashem, M.N.: "Determination of Coarse and Ultra-Fine Scale Compartmentalization by Downhole Fluid Analysis," paper SPE IPTC 10034 presented at the International Petroleum Technology Conference, Doha, Qatar (November 21–23, 2005).

Mullins, O.C., Rodgers, R.P., Weinheber, P., Klein, G.C., Venkataramanan, L., Andrews, A.B., and Marshall, A.G.: "Oil Reservoir Characterization via Crude Oil Analysis by Downhole Fluid Analysis in Oil Wells with Visible-Near-Infrared Spectroscopy and by Laboratory Analysis with Electrospray Ionization-Fourier Transform Ion Cyclotron Resonance Mass Spectroscopy," *Energy & Fuels* (2007) 21, No. 6, 2448–2456.

Mullins, O.C., Sheu, E.Y., Hammami, A., and Marshall, A.G. (eds.): *Asphaltenes, Heavy Oils, and Petroleomics*, New York, New York, USA, Springer (2007).

Raghuraman, B., Gustavson, G., Mullins, O.C., and Rabbito, P.: "Spectroscopic pH Measurement for High Temperatures, Pressures and Ionic Strength," *American Institute of Chemical Engineers Journal* (2006) 52, No. 9, 3257–3265.

Raghuraman, B., O'Keefe, M., Eriksen, K.O., Tau, L.A., Vikane, O., Gustavson, G., and Indo, K.: "Real-Time Downhole pH Measurement Using Optical Spectroscopy," paper SPE 93057 presented at the SPE International Symposium on Oilfield Chemistry, The Woodlands, Texas, USA (February 2–4, 2005).

Raghuraman, B., Xian, C., Carnegie, A., Lecerf, B., Stewart, L., Gustavson, G., Abdou, M.K, Hosani, A., Dawoud, A., Mahdi, A., and Ruefer, S.: "Downhole pH Measurement for WBM Contamination Monitoring and Transition Zone Characterization," paper SPE 95785 presented at the SPE Annual Technical Conference and Exhibition, Dallas, Texas, USA (October 9–12, 2005).

Ratulowski, J., Fuex, A.N., Westrich, J.T., and Sieler, J.J.: "Theoretical and Experimental Investigation of Isothermal Compositional Grading," *SPE Reservoir Evaluation & Engineering* (June 2003) 6, No. 3, 168–175; also presented as paper SPE 63084 at the SPE Annual Technical Conference and Exhibition, Dallas, Texas, USA (October 1–4, 2000).

Reynolds, A.D.: "Dimensions of Paralic Sandstone Bodies," *AAPG Bulletin* (1999) 83, No. 2, 211–229.

Shaw, J.M., and Zou, X.: "Phase Behavior of Heavy Oils," *Asphaltenes, Heavy Oils, and Petroleomics*, O.C. Mullins, E.Y. Sheu, A. Hammami, and A.G. Marshall (eds.), New York, New York, USA, Springer (2007), 489–510.

Sheu, E.Y., Long, Y., and Hamza, H.: "Asphaltene Self-Association and Precipitation in Solvents—AC Conductivity Measurements," *Asphaltenes, Heavy Oils, and Petroleomics*, O.C. Mullins, E.Y. Sheu, A. Hammami, and A.G. Marshall (eds.), New York, New York, USA, Springer (2007), 259–278.

Sirota, E.B., and Lin, M.Y.: "The Physical Behavior of Asphaltenes," *Energy & Fuels* (2007) 21, 2809–2815.

Stainforth, J.G.: "New Insights into Reservoir Filling and Mixing Processes," *Understanding Petroleum Reservoirs: Towards an Integrated Reservoir Engineering and Geochemical Approach*, J.M. Cubitt, W.A. England, and S.R. Larter (eds.), London, England, Geological Society of London SP 237 (2004), 115–132.

Tissot, B.P., and Welte, D.H.: *Petroleum Formation and Occurrence* (2nd ed.), Berlin, Germany, Springer-Verlag (1984).

Venkataramanan, L., Weinheber, P., Mullins, O.C., Andrews, A.B., and Gustavson, G.: "Pressure Gradients and Fluid Analysis as an Aid to Determining Reservoir Compartmentalization," *Transactions of the SPWLA 47th Annual Logging Symposium*, Veracruz, Mexico (June 4–7, 2006), paper S.

Whitson, C.H., and Belery, P.: "Compositional Gradients in Petroleum Reservoirs," paper SPE 28000, presented at the University of Tulsa Centennial Petroleum Engineering Symposium, Tulsa, Oklahoma, USA (August 29–31, 1994).

Wilhelms, A., and Larter, S.: "Shaken But Not Always Stirred. Impact of Petroleum Charge Mixing on Reservoir Geochemistry," *Understanding Petroleum Reservoirs: Towards an Integrated Reservoir Engineering and Geochemical Approach*, J.M. Cubitt, W.A. England, and S.R. Larter (eds.), London, England, Geological Society of London SP 237 (2004), 27–35.

Xu, L., Cai, J., Guo, S., Yi, P., Xiao, D., Dai, Y.D., Yang, S.K., Khong, K.C., Fujisawa, G., Dong, C., and Mullins, O.C.: "Real Time Carbon Dioxide Quantification Using Wireline Formation Tester to Optimize Completion and Drill Stem Testing Decisions," presented at the SPWLA Southeast Asia/Japan/Australia Technical Forum, Bangkok, Thailand (August 24–27, 2008).

Yudin, I.K., and Anisimov, M.A.: "Dynamic Light Scattering Monitoring of Asphaltene Aggregation in Crude Oils and Hydrocarbon Solutions," *Asphaltenes, Heavy Oils, and Petroleomics*, O.C. Mullins, E.Y. Sheu, A. Hammami, and A.G. Marshall (eds.), New York, New York, USA, Springer (2007), 439–468.

Zeng, H., Song, Y.-Q., Johnson, D.L., and Mullins, O.C.: "Critical Nanoaggregate Concentration of Asphaltenes by Low Frequency Conductivity," *Energy & Fuels* (in press).

Zhao, B., and Shaw, J.M.: "Composition and Size Distribution of Coherent Nanostructures in Athabasca Bitumen and Maya Crude Oil," *Energy & Fuels* (2007) 21, No. 5, 2795–2804.

CHAPTER TWO
The Photophysics of Reservoir Fluids:
The Scientific Foundation of Optical DFA Measurements

Introduction

Oil sample acquisition in open hole

Downhole fluid analysis grew out of sample identification requirements in openhole sample acquisition. First and foremost, whether a potential reservoir contains hydrocarbons or water or some other fluid such as CO_2 must be known. And because the production requirements for oil versus gas are quite distinct, identifying whether any hydrocarbon is oil or gas is essential. In addition, sample acquisition necessarily involves a reduction of pressure on the fluid, whereupon hydrocarbons can undergo a variety of phase transitions. Determining that the sample acquisition process does not yield a phase transition is thus also necessary. Oil wells are usually drilled with either water-base mud (WBM) or oil-base mud (OBM). The mud serves a variety of purposes, including pressurizing the borehole to prevent blowouts, preventing drilling fluid loss into permeable formations, lifting rock cuttings out of the well, lubricating drilling, and stabilizing the borehole. Some leakage of mud filtrate occurs through the mudcake into permeable zones. Depending on the fluids, the mud filtrate can be miscible with the formation fluids. Differentiating between mud filtrate and formation fluid is essential for the acquisition of representative samples. Fluid identification must occur at elevated temperatures and pressures. Typical borehole temperatures (thus tool operations temperatures) are limited by the decomposition temperatures of hydrocarbons over geologic time, roughly 175 degC. Using a worldwide mean thermal gradient of ¾ degC per 100 ft[1] gives a depth of roughly 20,000 ft. For a hydrostatic loading of ½ psi/ft, this depth corresponds to 10,000 psi. However, many formations are overpressured and subject to lithostatic loading of ~1 psi/ft, for which this depth corresponds to 20,000 psi. Fortunately, sample temporal variation is not very extreme; measurement frequencies of a few hertz often suffice. As is discussed shortly, the primary method of choice to accomplish these fluid identification tasks is visible–near-infrared (Vis-NIR) spectroscopy.

MDT Modular Formation Dynamics Tester

The industry leader in capturing representative samples in open hole is the Schlumberger MDT Modular Formation Dynamics Tester (Zimmerman, Pop, and Perkins, 1989, 1990). The current wireline DFA modules are components of the MDT tool. Figure 1 shows the formation sampler configuration and a photograph of the MDT probe. The probe establishes hydraulic communication with the formation to allow formation fluids to flow into the MDT tool. The MDT pump (Fig. 1) reduces the pressure in the flowline, causing formation fluids to enter the tool. In this process, formation samples can be acquired and stored in high-pressure sample bottles.

---

[1] *Oilfield customary units are an amalgam of English and metric units at best (see API gravity for the worst). This book just surrenders to the convention.*

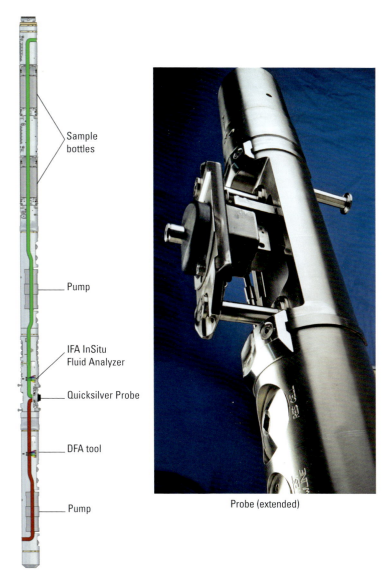

Figure 1. The MDT sampling tool descends into the well on a cable or wire after drilling is completed. The probe (a simple version shown in the photograph) is used to establish hydraulic communication with the fluids contained in permeable subsurface formations for the purpose of measuring pressure and extracting formation fluids for DFA and sample acquisition. Integrated modular spectrometers and other optics and measurement cells measure a variety of fluid properties. Also depicted are the pump and various sample bottles for storing the acquired formation fluids.

Vis-NIR spectroscopy of crude oil and water

Fluid identification

The primary measurements of DFA are based on Vis-NIR spectroscopy and were originally used for fluid identification, specifically oil, water, gas, and mud filtrate versus formation fluid. These measurements were selected for the following reasons:

- Vis-NIR requisite pathlengths for an optical density (OD) ~ 1 are ~1 mm, which is conducive to fluid analysis in the MDT flowline.
- NIR spectroscopy provides excellent distinction between oil and water, and the visible analysis distinguishes between OBM filtrate and crude oil.
- The optical hardware is compatible with downhole requirements of temperature, space, and power.

Figure 2 compares the spectra of several crude oils, an OBM filtrate, and water. The vertical axis plots OD. Because there is no light scattering, the OD simply equals the absorption of light. An OD of zero corresponds to no absorption, that is, transparency. The horizontal axis plots the wavelength of electromagnetic radiation, with the visible range where the human eye has sensitivity shown as a (vertical) rainbow. These colors are the response of the eye when irradiated by light of the corresponding wavelength. The NIR range is at longer wavelengths than that of visible light. The eye has no sensitivity here, but this radiation can be felt as heat.[2] There is quite a bit of spectral content in the NIR region for oilfield fluids of interest.

Water and oil are readily differentiated by virtue of their different absorption peaks in the NIR range (Safinya and Tarvin, 1991). For example, water has large absorption peaks at 1.45 um and 1.9 um. Interpretation algorithms can be constructed that are independent of the type of flow (slug, laminar, etc.) of the immiscible liquid phases (Hines, Wada, Garoff, et al., 1994). The OBM filtrate in Fig. 2 is colorless, whereas crude oils typically have strong coloration, and the coloration contrast of different crude oils extends into the NIR range. Coloration can be used for corresponding discrimination (Mullins, 1993; Mullins and Schroer, 2002). Note that coloration does not mean color or hue, such as red or blue. Crude oils are basically brown; thus, coloration means the extent of "brown" absorption. Gases are low density, so they are characterized by small optical absorption and no color. In addition, measurement of the optical index of refraction augments gas detection (Mullins, Hines, Niwa, et al., 1992). Corroborative measurements are desirable in such a remote location as a borehole. Retrograde dew is detected in large part by the use of fluorescence measurements (Mullins, Fujisawa, Dong, et al., 2006).

---

[2] The astronomer William Herschel discovered NIR in 1800 by its heating ability. After sending sunlight through a prism, to his surprise he found the greatest heat-inducing part of the spectrum was just beyond red visible light. It was the first demonstration that there is radiation invisible to the eye.

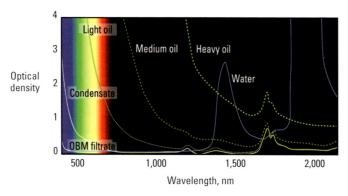

Figure 2. The different fluids of crude oils, OBM filtrate, and water can be spectrally resolved with the Vis-NIR spectra (Safinya and Tarvin, 1991; Mullins, 1993).

Sample cleanup to reduce miscible filtrate contamination is monitored by making measurements of the "color" of the flowline contents as a function of time. As the crude oil fraction of the flowline content increases, the color increases. The lack of a phase transition is monitored in part by the absence of gas.

Upon the realization that GOR could be measured downhole (Mullins, 1999; Dong, Hegeman, Mullins, *et al.*, 2005; Fujisawa, Mullins, Terabayashi, *et al.*, 2006), there was skepticism that this capability was needed, with the thinking that laboratory measurements would suffice. Specifically, laboratory reports on crude oils are often tens of pages loaded with numbers, so the question was posed repeatedly to this author, "What is the value of one number—GOR—determined downhole?" A notable exception to this skepticism is the paper by M. Hashem *et al.* anticipating the utility of simple fluid properties estimated in real time by correlation with OFA optical measurements (Hashem, Thomas, McNeil, *et al.*, 1997). It is always difficult to introduce a new product into a market that does not anticipate the value of that product. Thus, the GOR measurement was designed to address one of the most important concerns in MDT sample acquisition: miscible contamination, thereby making this measurement indispensible (Mullins, Beck, Cribbs, *et al.*, 2001; Dong, Mullins, Hegeman, *et al.*, 2002). OBM filtrates do not have asphaltenes whereas most crude oils do. This difference gives a significant color contrast between OBM filtrate and (most) crude oils. By monitoring the coloration of the flowline contents over time, an indicator of the cleanup process is provided (Fig. 3). In addition, OBM filtrates in the formation have hardly any dissolved gas whereas most crude oils have appreciable dissolved gas. Consequently, measurement of the dissolved methane content gives a second physics measurement for monitoring contamination, which had always been a vexing problem. Consequently, contamination monitoring provided the means to introduce downhole GOR as a successful service. The unanticipated success of DFA, which far exceeds sample contamination monitoring, is founded upon new understanding of the reservoir, as discussed in the first chapter.

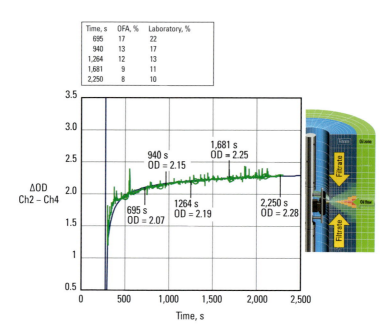

Figure 3. As fluid enters the wireline sampling tool set in a wellbore against a permeable formation (right), miscible filtrate contamination is initially present in significant concentration but is reduced with increased pumping time. The coloration of the fluid in the flowline increases with increasing crude oil concentration (left) because drilling fluid filtrate is usually very light in color compared with crude oil. Here, the quantitative prediction of contamination performed while pumping (Mullins and Schroer, 2002) agrees with subsequent laboratory analysis.

GOR, the first formal DFA product

The downhole determination of GOR was introduced to the field as a second method for monitoring cleanup (Mullins, Beck, Cribbs, *et al.*, 2001; Dong, Mullins, Hegeman, *et al.*, 2002). Because all clients want clean samples, GOR became an instant global success. After GOR was seen in real time, many new uses of this measurement became evident, and DFA was born (Fig. 4).

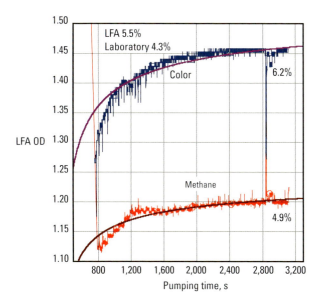

Figure 4. DFA was originally introduced to monitor contamination (Mullins, Beck, Cribbs, *et al.*, 2001; Dong, Mullins, Hegeman, *et al.*, 2002). The reduction of OBM filtrate contamination with pumping can be measured by using both color and dissolved methane content. In this figure, the methane signal was multiplied by 6 to put it on the same scale as color.

Both Figs. 3 and 4 list quantitative contamination measurements. To quantify contamination from the buildup curve, a semiempirical relation is used to compare cleanup with pumping time (presuming a constant pumping rate). For the case of no borehole—just a cylinder of contamination (mud filtrate) in virgin formation fluid—a simple geometric argument provides the desired relation. The MDT probe is a point sink in the middle of the cylinder (Fig. 5). With pumping, concentric spheres of fluid are drained into the sink probe. To drain a sphere at radius r, a fluid volume scaling with r^3 must have been drained first. The contamination in the spherical shell at radius r is given by the ratio of the intersection of the contamination cylinder with the spherical shell (essentially twice the fixed cross section of the cylinder) divided by the total surface area of the spherical shell, which scales as r^2. With pumping, the contamination is reduced by $r^{2/3}$, or for a constant pumping rate, the contamination is reduced by $t^{2/3}$ (Hammond, 1991).

Figure 5. Neglecting the borehole, the filtrate is depicted as a cylinder with the MDT probe indicated as a central point sink (Hammond, 2001). In the shell of fluid (light blue) at radius r, the filtrate decreases with r^2. To get to this radius, r^3 of fluid had to be drained. With constant pumping, the cleanup should scale with $t^{2/3}$.

In acquiring an oil sample in the presence of OBM filtrate, the oil coloration is not known a priori. At early times in the pumping process, significant OBM filtrate drains out of the near-wellbore region of the formation. After long pumping times, the OBM filtrate near the sampling point is pumped out and relatively clean crude oil enters the formation sampling tool. Coloration can be used to monitor the decreasing fraction of filtrate versus crude oil in the flowline (Figs. 3 and 4). Equation 1 gives the semiempirical universal curve that applies to the drainage of filtrate out of the formation presuming a constant pumping rate. This functional form is a modified result of the previously mentioned prediction of $t^{2/3}$ (Mullins and Schroer, 2000; Mullins, Schroer, and Beck, 2000):

$$OD_i(t) = OD_i(\infty) - \frac{c}{t^{5/12}}, \qquad (1)$$

where $OD_i(t)$ is the OD in one of the optical channels measuring the oil color at time t, $OD_i(\infty)$ is the OD obtained for an infinite pumping time (thus that for pure crude oil), c is a fitting coefficient, and t is pumping time. The buildup of coloration with pumping time is fit to Eq. 1 to give $OD_i(\infty)$ and c. The contamination is then the percent reduction of $OD_i(t)$ below $OD_i(\infty)$ (Mullins and Schroer, 2000; Mullins, Schroer, and Beck, 2000). For example, 10% filtrate contamination means that the measured color is 10% less than that of the pure crude oil. The real-time color measurements in Figs. 3 and 4 indicate the increasing fraction of crude oil in the flowline with increasing pumping time. The real-time log data and postjob laboratory data agree for the reduction of filtrate contamination with pumping time.

The field algorithm using Eq. 1 has garnered the sobriquet OCM* oil-base mud contamination monitoring.[3] Indeed, the OCM algorithm has been in use in the field for a long time. Variants of this algorithm continue to be explored.

At long pumping times, contamination enters the probe from around the backside of the borehole and vertically (Fig. 6). Because borehole effects are pronounced, the $t^{2/3}$ cleanup prediction is not generally realized. Modeling confirms that $t^{5/12}$ is the most probable power of time; moreover, the unfavorable geometry introduced by the borehole causes cleanup to take much longer than predicted by $t^{2/3}$ (Alpak, Elshahawi, Hashem, et al., 2006).

---

[3] This acronym is the author's initials, leading to confusion over its origin, a confusion strongly promulgated by the author.

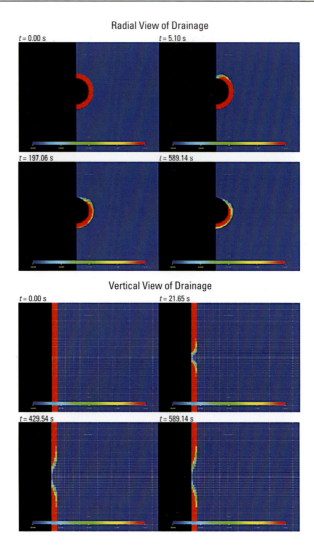

Figure 6. Modeled radial and vertical views of cleanup with a single probe show contamination (red) is in the peripheral part of the flow entering the probe (Alpak, Elshahawi, Hashem, *et al.*, 2006).

Contamination enters the probe peripherally both vertically and from the sides of the borehole while the formation fluid enters more radially into the probe (Fig. 6). Thus, focused fluid acquisition can result in much faster cleanup (Hrametz, Gardner, Waid, et al., 2001). The Quicksilver Probe* technology exploits these principles to efficiently separate contaminated fluids from pure formation fluid. Two probe areas are arranged concentrically, with each probe connected to an independent pump and separate flowline (Fig. 7) (Del Campo, Dong, Vasques, et al., 2006; O'Keefe, Eriksen, Williams, et al., 2006; Weinheber and Vasques, 2006). The inner extraction area is isolated from the surrounding "guard" area around the perimeter, which collects the contaminated flow coming from the sides, top, and bottom of the flow. Quicksilver Probe focused fluid extraction is similar to the electrical analog of focusing current into the formation with guard electrodes.

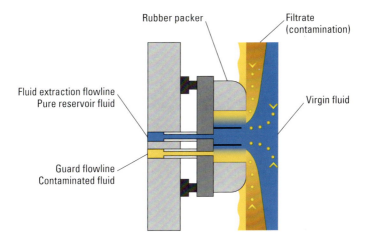

Figure 7. Focused fluid extraction is achieved by the two concentric probes of the Quicksilver Probe tool. The inner probe collects virgin formation fluid to a flowline while the outer guard probe collects contaminated fluid. After Del Campo, Dong, Vasques, et al. (2006). © 2006 Society of Petroleum Engineers.

The quantification of filtrate contamination during sample acquisition is a major concern only partially alleviated by the OCM algorithm. Nevertheless, ground truth is often difficult to establish. Laboratory methods to quantify contamination typically use the "skimming technique," in which gas chromatography (GC) is run for the dead, contaminated oil and on the mud filtrate. GC peaks in excess of the norm for the oil in the appropriate carbon number range are considered contamination and thereby quantified. The problem, of course, is knowing what the norm is. If the OBM filtrate is a diesel-base fluid, then laboratory determinations of contamination have significant error bars. In addition, valid samples of the mud filtrate can be difficult to obtain for jobs performed in remote locations. Both downhole and laboratory determinations of contamination have error bars—if discrepancies occur, then careful analysis is mandated. Indeed, for all downhole and laboratory measurements of fluid properties, there should be no presumption of accuracy for laboratory measurements or for downhole. Analytical accuracy is addressed in detail in the subsequent discussion of the chain of custody treatment of samples.

Oil chemistry

The purpose of oilfield sample acquisition is defined by chemical considerations. The economics of the production of hydrocarbons depends critically on chemical composition. Crude oils and gases vary enormously in their properties, ranging continuously from tar to dry natural gas (Fig. 8). Correspondingly, the GOR varies from nearly zero to infinity for reservoir hydrocarbons. The GOR is a strong determinant of specifications for production facilities and production strategy among other critical issues such as economic value. The mass fraction of asphaltene varies from 0% to 15% in hydrocarbon deposits, with Athabasca bitumen at the upper end with its corresponding huge and temperature-dependent viscosity. Asphaltenes, colloidally suspended solids in crude oil, have an enormous impact on all phases of production, transportation, and refining. Organic acids can dramatically alter emulsion stability and rock wettability of a crude oil. In addition, the nonhydrocarbon components of a crude oil, such as CO_2 and H_2S, can have an inordinate impact on facilities metallurgy, safety, and corrosion in addition to reducing economic value.

Figure 8. A wide range of hydrocarbons (here at low pressure) occurs in subsurface formations. Varying and often large quantities of dissolved gas are in the reservoir liquids. Every aspect of hydrocarbon production and utilization, from economics to facilities design to production strategies, depends on the types of produced fluids.

The classification of crude oils in Table 1 incorporates both live and corresponding dead crude oil samples.[4]

Table 1. Classification of Oil and Gas (McCain, 1990)

Oil Type	GOR (live oil), ft^3/bbl	API Gravity (dead oil)	Color of Stock-Tank Oil
Black oil	<2,000	<45	Very dark colored to black
Volatile oil	2,000–3,300	>40	Colored, usually brown
Retrograde condensate	3,300–50,000	50–60	Lightly colored
Wet gas	>50,000	>50	Clear
Dry gas	∞	No liquid	No liquid

In addition to this classification, there is heavy oil to be considered (Table 2).

Table 2. Classification of Heavy Oil

Oil Type	Viscosity, cP	API Gravity
Conventional crude oil	<10,000	>20
Heavy oil	>10,000	10–20
Extraheavy oil	–	<10

Crude oils can undergo a variety of phase transitions, producing gas, dew, asphaltenes, wax, gas hydrate, and diamonoids. The various solids associated with hydrocarbon phase transitions are obviously problematic (Fig. 9). These solids fall under the banner of flow assurance and are of particular concern in deepwater, where there is limited access to seafloor pipelines and facilities, let alone well completions.

---

[4] *Live crude oils contain dissolved gases, as under reservoir conditions, whereas dead crude oils have been depressurized and their gases lost.*

Figure 9. Various organic solids and inorganic solids (only one of the latter is shown here) are of concern in the production of oil and gas. All these solids fall under the banner of flow assurance. From Mullins, Rodgers, Weinheber, *et al.* (2006). © 2006 American Chemical Society.

To evaluate the myriad of concerns regarding the production of oil and gas, chemical analysis of the sample is of paramount importance; sample acquisition for evaluation is one of the most important motivations for drilling evaluation wells. The relatively new technology of DFA introduces a real-time approach to the chemical evaluation of oilfield samples. The technology of DFA is unique in the upstream oil business, and the historical context of the evolution of DFA is instructive regarding DFA technology selection. This section reviews the origins of DFA and the scientific foundation of the most common DFA practices to date. The reader should note that this is an evolving and expanding story. New technologies are being brought to bear in DFA. Nevertheless, the review here is not likely to be outmoded any time soon; rather, new methods will supplement, not supplant, the methods described. Specifically, the dominant role of optics and optical spectroscopy is likely to remain a fixture for DFA.

Analytical chemistry

Analytical chemistry has two primary methodologies for analyzing complex mixtures such as crude oil: separation of the mixture into its constituent components and bulk spectroscopy analysis.

In separation science, some method is used to isolate individual components of the mixture. Examples include chromatography such as GC, liquid chromatography, and size-exclusion chromatography. Electrophoresis is another separation method used extensively for biological systems. Mass spectroscopy is a separation method broadly employed in many disciplines. For oilfield samples, GC is routinely used and gives reasonable resolution for the light ends. The familiar SARA analysis separates the crude oil into classes of compounds only obliquely related to chemical characterization: saturates, aromatics, resins, and asphaltenes. SARA analysis is routinely employed for crude oils even though it is of marginal utility for any but the most rudimentary purposes. The shortcomings of these separation approaches are typified by SARA analysis, for which the characterization of >50,000 components in a typical (dead) crude oil into the four pseudocomponents of saturates, aromatics, resins, and asphaltenes mandates a clearly and properly defined objective if it is to be useful. The foundation of petroleomics is to identify each of the huge number of chemical constituents of crude oil. This is largely enabled in a laboratory setting and is treated elsewhere (Mullins, Sheu, Hammami, *et al.*, 2007).

The other primary method employed in analytical chemistry is bulk spectroscopy. The entire sample of a complex mixture can be subjected to a radiation field and the response measured. Typical radiation for interrogation includes any of the numerous electromagnetic spectral ranges such as X-ray, ultraviolet (UV), visible, infrared, and microwave. Other distinct fields include acoustics and various particles such as neutrons or electrons. Typically, the attenuation of transmission is measured as a function of the frequency or energy of the incident radiation. For example, the absorption of light can be measured versus wavelength to give an optical absorption spectrum. Detection of constituents in the mixture follows from the spectral profile. Nonlinear effects on the spectrum, if any, of mixing components must be determined to perform the inverse analysis of complex spectra and obtain constituent concentrations from the spectrum of the mixture. Spectral analysis can suffer from a uniqueness problem in that different chemical moieties can produce unresolvable spectral features. In such cases, pseudocomponents or some form of lumping must be performed.

For complex mixtures such as crude oils, there is no question that separation science provides much a higher resolution of constituents and is preferred if detailed compositional analysis is needed. However, for monitoring the contents of a flow stream, spectral analysis possesses intrinsic advantages. Spectral analysis can often be performed over the entire cross section of the flow stream whereas separation methods normally sample only a small fraction of the total sample volume of interest. Moreover, separation science cannot be performed continuously over time on a sample flow stream but is performed in batch mode on a small sample, and higher resolution takes more time. For a flow stream, this means that most of the flow stream contents remain unanalyzed. In contrast, spectroscopy can measure continuously, limited only by the bandwidth of the measurement. For a large bandwidth, virtually all the contents of a flowline are monitored. For monitoring processes in which the flowline contents are known to fall within a limited range, bulk spectral measurements are often preferred. Moreover, spectral measurements are usually much simpler. For example, no processing of the flowline sample is generally required in spectroscopy, but it is for separation science. For downhole sample acquisition, spectroscopy is far simpler, less expensive, and thus more likely to be used than separation science. In addition, in the process of downhole sample acquisition, the flowline contents of the MDT tool share similar features to process flow streams in refineries, where spectroscopy is popular. For example, the

Btu content of a natural gas flow stream can be continuously monitored by using NIR spectroscopy (Brown and Lo, 1993). Even though the range of potential fluids and conditions is rather large in downhole sample acquisition, spectroscopy is preferred over separation science for monitoring the sampling process. Moreover, for major DFA objectives such as detecting compartments or connectivity, the key fluid signatures of note are often basic, such as GOR or asphaltene content. In these cases, spectral analysis is preferred because the high-resolution separation methods are less sensitive to changes in composition over a very broad range of components. That is, high resolution tends to preclude analysis over a broad range. For DFA purposes, spectral methods have established the great value of performing analytical chemistry in the downhole environment, and the possible value of separation science can be revisited.

Vis-NIR spectroscopy is the foundation of DFA measurements. Although common in the downstream oil business, optical spectroscopy is new to the upstream oil business. The upstream and downstream branches of the oil business are not in routine communication, to put it mildly. Consequently, a technology that is common practice in one branch of the industry can be totally unfamiliar to the other. Moreover, DFA spectral measurements consist of both the visible spectral range and the NIR spectral range. This combination is not common for monitoring process streams in any industry; spectroscopy equipment vendors have no standard products that combine these spectral ranges for field applications such as refineries. Consequently, a review is in order of the relevant considerations and basic scientific principles that govern DFA spectroscopy measurements. In addition, the DFA suite of measurements includes fluorescence measurements. Fluorescence measurements have been employed especially for particular phenomenological objectives in the upstream oil business. For example, mud loggers routinely use fluorescence methods to check for oil shows on rock cuttings during drilling. Nevertheless, establishing the scientific foundations of fluorescence measurements has not been the purview of mud loggers, again to put it mildly. Indeed, it is incumbent upon us to develop the scientific foundations of commercial DFA services, which we have done, and to provide a review, which is in this chapter.

Molecular spectroscopy

Molecular energy levels

Molecules possess three primary energetic modes (in addition to translation): electronic, vibrational, and rotational (Atkins and Friedman, 2005). Transitions between different energy states in the molecule originate from various processes, such as molecular collision or radiative absorption or emission. Photoexcitation is of particular concern for DFA. All the different molecular modes are quantized, so excitation by photoabsorption occurs only when the photon energy matches a quantized transition within the molecule. This is mandated by the law of conservation of energy. Other conservation laws also apply to molecular excitation processes relating to symmetry and the conservation of angular momentum.

For a mixture of molecules, the sum of the absorption spectra of the different chemical constituents generally gives the resulting spectrum of the mixture. This presumes that intermolecular interactions do not affect the spectra, which largely applies to crude oils and alkanes. These three energetic modes of molecules are quite distinct in their corresponding energies. Figure 10 shows the wavelengths (on a log scale) of electromagnetic radiation over 13 orders of magnitude, along with the corresponding molecular transitions of interest.[5] The wavelength of radiation is inversely related to frequency:

$$\lambda \nu = c, \qquad (2)$$

where λ is the wavelength, ν is the corresponding frequency, and c is the speed of light (3×10^{10} cm/s). The unit of frequency is cycles per sec or hertz (Hz). Because the frequencies are so large, frequency is often given as ν/c (or $1/\lambda$) with units of wavenumber in cm^{-1}. A photon of 1-um wavelength is 10,000 cm^{-1}. Photons are the quantized "particles" of light, and the energy E of a photon is

$$E = h\nu, \qquad (3)$$

where h is Planck's constant. There is unit probability per electron of excitation of a molecule when integrating over all frequencies. That is, the total excitation probability per electron is a conserved quantity. Consequently, it is convenient to discuss the probability of different molecular transitions in terms of oscillator strength, and the sum of the oscillator strength for all transitions of an electron is one.

Electronic transitions of interest correspond to excitation of the outermost electrons (valence electrons) to excited states. These electronic transitions correspond to the UV and visible spectral ranges. The oscillator strength of valence shell electronic transitions is of order 1, and these transitions correspond to very strong absorption of light. The colors of the visual field are dominated by electronic transitions. For crude oils, the visible color corresponds to the spectral edge of electronic transitions, which is nowhere near the center of the electronic absorption at much shorter wavelengths. More accurately, crude oils consist of a mixture of huge numbers of chromophores, which are light-absorbing molecules. Most crude oil chromophores absorb in the UV, not the visible. Consequently, relatively small electronic absorbances (~1 OD/mm) are recorded for crude oils in the Vis-NIR spectral range. Just as electronic absorption is strong, so is emission from electronic transitions. For example, visible-light sources that operate on the principle of electronic emission include mercury lamps, sodium lamps, and light emitting diodes. These are

---

[5] *The visible spectral range is where the human eye has sensitivity and is approximately 400 nm to 700 nm. This narrow spectral range coincides with the similarly narrow spectral range where water is transparent and also where the sun has its maximum output.*

all quite efficient at converting electrical energy to optical energy. Figure 10 also shows the X-ray range, which corresponds to the excitation of inner shell electrons that are closer to the nucleus and require more energy for excitation. X-ray spectroscopy has been useful for the characterization of carbon (Bergmann, Groenzin, Mullins, *et al.,* 2003), sulfur (George and Gorbaty, 1989), and nitrogen (Mitra-Kirtley, Mullins, van Elp, *et al.,* 1983) in asphaltenes, where other methods are not as effective. Gamma rays (Fig. 10) correspond to nuclear transitions and are used in a variety of nuclear logging applications (Ellis and Singer, 2007).

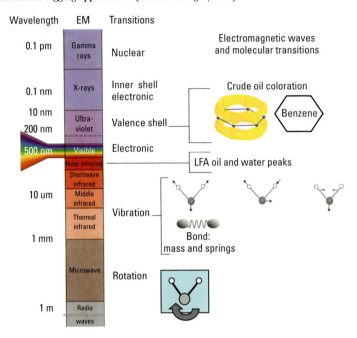

Figure 10. The electromagnetic (EM) spectrum and corresponding molecular transitions of comparable energy are shown.

Fundamental modes of vibrational transitions are excited by the mid-infrared (MIR) spectral range (25-um to 2.5-um wavelength). These bands have an oscillator strength of ~10^{-2}, so they are much weaker than electronic transitions but still fairly strong. The NIR spectral range corresponds to overtones and combination bands of vibrational modes. That is, a single photon can excite two quanta of molecular vibrations; if both are the same vibrational mode, the transition is an overtone, and if they are different vibrational modes, the transition is a combination band. The NIR bands are called forbidden bands because they should have zero oscillator strength to the zeroth order. "Imperfections" in the absorption process and in molecular vibrations cause these absorption bands to be nonzero, but their oscillator strength of $\sim 10^{-4}$ is quite weak. Consequently, long pathlengths (millimeter range) are needed to achieve significant absorption. This concept of overtones is familiar from a musical setting. For example, the tonal differences of different musical instruments are strongly dependent on their overtones. It is common for guitar players to suppress the fundamental mode on a guitar string to excite only overtones.

Molecular rotational transitions occur in the microwave range, but essentially for the gas phase. In liquids, microwaves correspond to some degree of rotational collective modes and are more complicated than simple rotation. These are also fairly weak transitions and are currently not of interest for DFA. Microwave ovens operate at a frequency of 2.45 GHz and heat food primarily by exciting water. It takes centimeters of food to attenuate the microwave radiation in large measure because microwave ovens are tuned away from the giant dipole resonance of liquid water at 17 GHz. If microwave ovens operated at 17 GHz, they would cost a lot more and the absorption of microwaves would be so strong that food would char on the outside while remaining raw inside. A common misconception about microwaves is that they cook "inside out."

As discussed shortly, the Vis-NIR spectral range is of interest for DFA. What about the corresponding hardware for this spectral range? The requisite Vis-NIR spectral range is quite broad, so single-frequency line emission is not so suitable. Instead, an incandescent source, similar to a standard light bulb, is appropriate. Filaments of incandescent lamps generate light by virtue of being blackbody radiators; objects that are hot emit radiation.[6] For blackbody radiation hotter objects emit much more light and the emitted radiation per unit area of surface varies as T^4 (Stefan's law, where T is temperature) (Jenkins and White, 2001). The tungsten-halogen lamp is the source of choice for the Vis-NIR spectral range and operates at $\sim 3{,}000$ K, producing a maximum emission λ_{max} at ~ 1 um wavelength. Wein's displacement law shows the inverse relation between λ_{max} and T, where λ_{max} (in microns) = 2,897/T (K). The Vis-NIR range can be conveniently piped in standard fiber optics. Only glassy substances such as silica can be pulled to make fiber optics. However, there is no convenient glassy substance with MIR transmission. Synthetic sapphire is transparent throughout the Vis-NIR range and is strong and tough; sapphire windows are capable of holding off large pressures. Si photodiode detectors, with a 1.1-um band-gap, work even when hot. InGaAs photodiode detectors, with a 1.7-um band-gap, have much higher thermal noise when hot but still provide usable signals. Extended (smaller band-gap) InGaAs detectors

---

[6] *The adjective "black" seems to be a misnomer for a good radiator. However, objects that are good emitters at some wavelength must also be equally good absorbers at that wavelength. Otherwise, two objects of the same temperature and emission strength but differing absorption strength could be placed side by side with a net transfer of energy, thus heat, from one body to the other with a concomitant growing temperature difference. This would violate the second law of thermodynamics.*

can also be used for long-wavelength detection. For typical spectrometers, absorbances of 1 (10% transmission) are readily measured, but absorbances much higher than 2 (1% transmission) can exhibit nonlinearities. Likewise, very small absorbances are not easily measured. Pathlengths of 2 mm are convenient for a downhole flowline and often yield roughly 1% to 10% light transmission for typical samples.

In a remarkable coincidence, the MIR and NIR spectral ranges differ both in molecular transitions and in hardware. The MIR range corresponds to fundamental vibrational bands and also necessitates different optical trains than the NIR range. The strength of the MIR bands necessitates ~10 um of pathlength, which is too small for facile downhole applications with the possible exception of total internal reflection measurements. In addition, MIR sources are low brightness, and Wein's displacement law stipulates the use of lower temperature incandescent sources for λ_{max} in the MIR range. Stefan's law dictates that these temperature sources are much less bright than NIR sources. Moreover, as mentioned previously, there are no standard MIR fiber optics. MIR window materials are problematic, and MIR detectors are sensitive to thermal noise because MIR radiation is basically thermal in hot environments. NIR is far more useful downhole than MIR.

Basic optical principles in spectroscopy

The prototypical starting point for spectroscopy is an optical beam or a "light" beam of intensity I and wavelength λ incident on a very thin sample (Fig. 11). Light generally refers to visible radiation, but here any electromagnetic wavelength range is implied. For example, the same reasoning applies to X-ray beams.

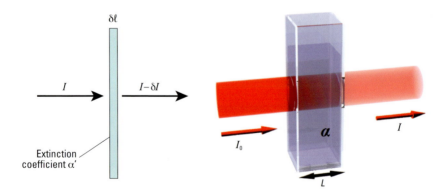

Figure 11. Left: Light absorption in a very thin sample is characterized by an extinction coefficient α'. Right: For a thick sample, the attenuation of light is exponential.

As the incident beam of intensity I traverses the very thin sample of thickness $\delta\ell$, the extinction coefficient of the sample for light of wavelength λ is α'. The reduction of intensity δI (thus the negative sign) of the optical beam after traversing the sample is (Jenkins and White, 2001)

$$\delta I = - I\alpha'\delta\ell. \qquad (4)$$

The reduction of intensity is proportional to the intensity. This is true only for very thin samples where the intensity reduction δI is small compared with the intensity. This is a first-order differential equation. For thick samples, the integral of Eq. 4 is needed:

$$\int_{I_0}^{I} \frac{\delta I}{I} = -\alpha' \int_{0}^{L} \delta\ell, \qquad (5)$$

where I_0 is the incident intensity, I is the intensity after traversing the sample, and L is the sample thickness (for the thick sample). The sample is presumed to be homogeneous, with a uniform optical extinction coefficient α' (homogeneous on the scale equal to or larger than the wavelength of light λ). Thus, α' can be placed outside the integral because it has no dependence on position ℓ in the sample (nor any lateral dependence). If the sample is not homogeneous, then the extinction coefficient α' depends on position in the sample and the analysis becomes much more complex. The solution of Eq. 5 is

$$\ln \frac{I}{I_0} = -\alpha' L$$

or equivalently

$$\frac{I}{I_0} = e^{-\alpha' L}. \qquad (6)$$

I/I_0 is defined as the optical transmission. The proportionality of the reduction of intensity to the intensity (Eq. 4) for a thin slice of sample has the consequence that the reduction of intensity through a thick sample is exponential. The same equation applies for radioactive decay by substituting time for pathlength. For radioactive sources, there is an exponential decay of the source strength with time.

Equation 6 shows that the intensity is not linear in pathlength or in extinction coefficient. Rather, the natural log of the intensity ratio is linear in pathlength and extinction coefficient. It is more useful to use log base 10 rather than the natural log (log base e). Exploiting a property of logarithms, Eq. 6 gives

$$\log_{10}\left(\frac{I}{I_0}\right) = -\alpha L, \quad (7)$$

where $\alpha = \alpha' \log 10(e)$ and e is Euler's number (2.71828...). In Eqs. 4 through 7, α is a property of the material, such as a pure element of a certain crystal structure, and is a function of the wavelength of light. However, if the constituents of the homogeneous sample can change, such as a crude oil, then it is often more useful to express the extinction coefficient in terms of the variable chemical composition of the sample. Equation 7 can be rewritten as the Beer-Lambert law:

$$\log_{10}\left(\frac{I}{I_0}\right) = \sum_i \varepsilon_i c_i L \equiv A, \quad (8)$$

where ε_i is the molar absorptivity of component i and c_i is its concentration in the sample. The molar absorptivity ε_i for component i is a function of wavelength. A in Eq. 8 is defined as the absorbance (if there is a presumption of no light scattering). The total absorbance is equal to the sum of the absorbances from the individual chemical components. Spectral analysis of a chemical mixture can be resolved into the sum of the spectra of its constituents, provided there are no nonlinear chemical interactions. Hydrocarbons are "ideal" in their chemical behavior in that they have very weak intermolecular interactions. Thus, crude oil spectra exhibit linear behavior. For DFA purposes, the absorption spectra of hydrocarbons are linear and additive, so no spectral nonlinearities need to be considered (Mullins, Joshi, Groenzin, et al., 2000). The only exception to this rule is low-pressure gases. At low pressures, collisional line broadening is reduced and spectral line widths can become very narrow. Measuring narrow line widths with filters of larger bandwidth yields nonlinearities in the absorption spectra. For methane, linear narrowing is evident below 1,000 psi.

If there is also optical light scattering, then this is a second mechanism that attenuates the optical beam. Absorption is the process in which photons of light are absorbed whereas scattering is the process in which photons of light are reflected or refracted in a new direction. Things that are black absorb (visible) light and things that are white scatter light. For light scattering, the total intensity of light is not diminished; however, the intensity of light in the direction of the initial light beam is reduced. Experimentally, it is appropriate to consider scattering as reducing the measured beam intensity because for most spectral instruments, including all DFA optical tools, the light collection optics accept only a very small angular and area aperture corresponding to accepting the initial optical beam. Thus, scattering reduces measured optical throughput.

The combined effects of absorption and scattering are included in the term "optical density." In general, nonlinear terms can arise that depend on both scattering and absorption effects. For example, multiple scattering of single photons can result in the collection and measurement of photons that traverse a longer circuitous path through the sample. Nevertheless, these nonlinear effects are usually small, and the OD is then the sum of the absorption plus light scattering:

$$OD = -\log_{10}\left(\frac{I}{I_0}\right) = absorption + scattering$$

or equivalently

$$\frac{I}{I_0} = 10^{-OD}. \tag{9}$$

Scattering

Scattering is caused by an impedance contrast associated with phase boundaries, density differences, and concentration differences. Colloidal systems are often excellent scatterers of light. Colloidal systems are defined as having a discrete phase in a continuous phase where the particles are in the size range ~2 nm to 200 nm. Table 3 shows a classification of colloidal systems that can yield light scattering.

Table 3. Classification and Examples of Light-Scattering Colloidal Systems

Continuous Phase	Dispersed Phase		
	Gas	Liquid	Solid
Gas	None[†]	Liquid aerosol (fog, clouds, steam)	Solid aerosol (smoke, particulate matter in air)
Liquid	Foam (shaving cream)	Emulsion (milk, mayonnaise, hand lotion)	Sol (blood, pigmented ink)
Solid	Solid foam (aerogel, Styrofoam®, pumice, Ivory® soap)	Gel (gelatin, jelly, cheese, opal, quicksand)	Solid sol (cranberry glass)

[†] Gases are mutually miscible.

In colloidal systems, a property which almost becomes defining is that the interfacial surface area is huge and surface effects can become dominant. For example, interfacial areas of hundreds of square meters per cubic centimeter are not uncommon for colloidal systems. Enormous surface area occurs even if only one dimension of the dispersed phase is in the colloidal range of 2 nm to 200 nm. For example, foams are characterized as colloid even though they often have only one dimension in the colloidal range, the film thickness. Consequently, there are lamellar colloids, fibrous colloids, and corpuscular colloids.

Asphaltenes, a solid phase, are colloidally suspended in crude oil as nanocolloids (~2-nm particle size) (Mullins, Betancourt, Cribbs, *et al.*, 2007; Betancourt, Ventura, Pomerantz, *et al.*, in press). However, due to the small size and small impedance contrast, the nanocolloid suspension of asphaltenes does not scatter light (Mullins, 1990). If the asphaltenes are destabilized, then they aggregate into large particles that are very efficient light scatterers. Asphaltenes are surfactant in oil/water and oil/rock interfaces, which expands their interfacial importance beyond their colloidal dispersion in oil.

Indeed, the length scale of the scattering object versus the wavelength of interest is of paramount importance in determining the characteristics of the scattering process. For example, scattering can be either wavelength dependent or wavelength independent. For particles that are large compared with the wavelength of light, the scattering is wavelength independent and large in cross section (provided that the impedance contrast is large). On the other hand, for particles (of size r) that are comparable to or smaller than the wavelength of light λ, the scattering is wavelength dependent and generally the scattering becomes much weaker. For $r/\lambda \ll 1$, the familiar Rayleigh scattering results and the scattering cross section varies as $(r/\lambda)^4$, less than its geometric cross section πr^2.

In addition, Rayleigh scattering produces strongly polarized scattered light. Snow and clouds appear white as a result of wavelength-independent light scattering because both snow and clouds are composed of particulates that are much larger than 1 um (Fig. 12). In addition, incoherent multiple scattering events reflect all incident light. In contrast, the sky is blue; light scatters from density fluctuations in air that are much smaller than a micron, which is roughly the wavelength of light. This weak scattering is Rayleigh scattering, and blue light is preferentially scattered. For example, as sunlight traverses more atmosphere at sunrise and sunset, more blue light is scattered out, and the sun appears quite red. Cigarette smoke takes on a more blue hue after inhalation and exhalation; the larger particles in cigarette smoke are trapped in the lungs and airways, and the remaining small particles in exhaled cigarette smoke produce wavelength-dependent light scattering.

For DFA purposes, the wavelength of light is ~1 um. Thus, sand, particles of mudcake or mud, and emulsion droplets are strong, wavelength-independent scatterers of light. Disbursed clays can be wavelength-dependent scatterers with somewhat lower scattering magnitudes.

Figure 12. Light scattering from snow and clouds is wavelength independent, but light scattering from the sky is wavelength dependent.

For pedagogic purposes, consider a familiar example of long wavelengths exhibiting minimal scattering from relatively small objects: the scattering of water waves from the pilings of a pier. If short-wavelength water waves are created near a piling of much bigger dimension, the waves that strike the piling all reflect back. However, as we have all seen (Fig. 13), longer wavelength waves of water pass through as if the pilings were not there.

Figure 13. Long-wavelength waves traverse through pilings of much shorter lateral dimension, whereas short-wavelength waves would efficiently reflect (scatter) from the pilings. The same principles apply to optical radiation.

Figure 14 shows log data from the OFA* Optical Fluid Analyzer (the predecessor of the LFA Live Fluid Analyzer module for the MDT tool) for sample acquisition (Mullins and Schroer, 2000). The huge light scattering evident in the displayed channels can be subtracted out to give a clean coloration buildup curve. The subtraction works because the scattering is wavelength independent. (In addition, the OD is not too high, so no nonlinear effects in detector response occur.) In principle, to obtain the coloration buildup curve of Channel 3, any longer wavelength channel could be used for subtraction. However, typically channels close in wavelength are chosen because they are more likely to be the exact same type of detector, and any effects from wavelength dependence in the scattering are mitigated. Spectral subtraction to eliminate or reduce the effects of light scattering is performed on every job involving any optical tool.

In contrast, Fig. 15 shows wavelength-dependent light scattering in log data from a high-permeability zone. The mud was not designed to bridge the large pore throats (to avoid thick mudcake that might lead to swabbing the well). Consequently, the fine mud solids that loaded into the formation had to be removed during MDT sampling (Mullins, Schroer, and Beck, 2000). The fine solids cleaned up after 20 min, after which a clean coloration buildup curve was obtained. Analysis of the buildup curve reveals the contamination. The particulates clean up faster than the filtrate contamination for two reasons. The mud fines enter the formation in radial flow but are removed in spherical flow. They have a finite penetration length and eventually hang up in a throat. After some pumping time, the corresponding location of the mud fines from the probe exceeds their penetration length. In addition, the mud fines require some minimum energetic flow velocity to mobilize. The flow velocity drops off away from the probe, limiting the mobility of fines in the far field.

Asphaltenes show both types of scattering dependent on the flocculation condition (Fig. 16) (Joshi, Mullins, Jamaluddin, et al., 2001). For the live crude oil in Fig. 16, the onset pressure is 8,200 psi. In the top plot, the pressure on the live crude oil was reduced slightly below the asphaltene onset pressure (by ~200 psi). In contrast, for the bottom plot, the pressure was reduced much below onset (by ~2,000 psi). The top plot shows small, wavelength-dependent light scattering. Moreover, the sedimentation time is long. The particles are small, remain suspended, and continue to scatter light at long times.[7] The bottom plot shows large, wavelength-independent light scattering. The sedimentation rate is large because the particles are large, so the scattering diminishes rapidly in time. The laboratory industry standard to detect asphaltene flocculation is to measure the transmission of light at 1,600 nm, which is shown in Fig. 16 to be maximally transmissive. This wavelength is chosen in the laboratory so that cells of larger pathlength can be used (Hammami and Ratulowski, 2007).

---

[7] *The sedimentation equation for a particle of radius r and density contrast $\Delta \rho$ is determined by buoyancy forces counteracted by viscous drag. The viscous drag force is the product of particle velocity v and the Stokes viscous drag coefficient β, where $\beta = 6 \pi r \eta$, and η is viscosity. The sedimentation rate is $v = 2r^2 \Delta \rho g / 9 \eta$. The factor g is the Earth's gravitational acceleration.*

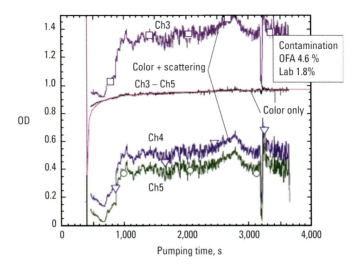

Figure 14. The large wavelength-independent light scattering in the OFA data in Channel 3 can be subtracted out, yielding a clean coloration buildup curve (Ch3 − Ch5) (Mullins and Schroer, 2000). © Society of Petroleum Engineers.

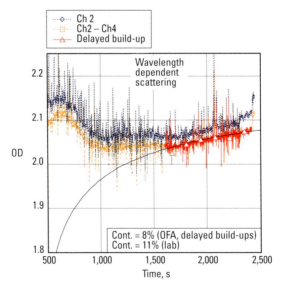

Figure 15. Wavelength-dependent light scattering could not be subtracted out (Mullins, Schroer, and Beck, 2000). The magnitude of the wavelength-dependent scattering is estimated by the difference between the optical data and the fitting curve (black). This type of scattering occurred because mud fines had loaded into the formation and flowed back into the MDT tool upon sampling. The fines cleaned up after 20 min, and subsequently a coloration buildup curve was obtained.

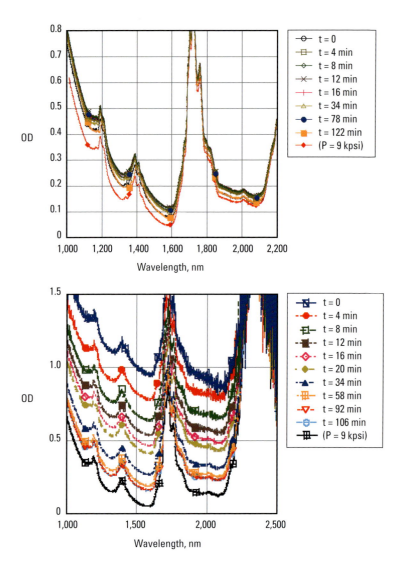

Figure 16. In both spectral plots of asphaltene flocculation, the smallest OD spectrum is obtained above asphaltene onset pressure (no scattering), with all other spectra recorded as a function of time at a pressure below asphaltene onset. In the top plot at pressure slightly below onset, small asphaltene flocs are formed, scattering is small and wavelength dependent, and the sedimentation rate is small. In the bottom plot for a pressure much below onset, large asphaltene flocs are formed, scattering is large and wavelength independent, and the sedimentation is fast. From Joshi, Mullins, Jamaluddin, *et al.* (2001). © 2001 American Chemical Society.

It would be very desirable to detect asphaltene onset pressure downhole. However, light scattering is not uniquely produced by asphaltene flocculation. Mud solids, emulsions, sand, clay, and other sundry particulates in flow can yield light scattering. However, there is a property of asphaltene flocculation that may assist in uniquely identifying the source of light scattering that results from asphaltenes. Figure 17 shows spectra of a live crude oil collected above (red curve) and below (blue curve) the asphaltene onset pressure. Large light scattering is evident in the spectrum acquired below the onset pressure (Joshi, Mullins, Jamaluddin, *et al.*, 2001).

An additional spectral trace is shown in Fig. 17. The Cary 5 UV-VIS-NIR spectrometer that was used to acquire this data collects a spectrum by sweeping the wavelength from long to short over a 4-min acquisition time for these spectra. That is, at each wavelength, optical transmission data is acquired for a selectable period of time, then a stepping motor advances the grating position to the next shorter wavelength, and again optical transmission data is acquired.

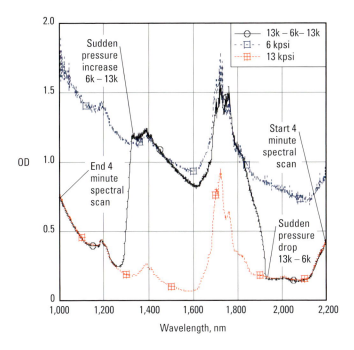

Figure 17. In addition to the spectra of a live crude oil at a pressure above asphaltene onset (red curve) and below asphaltene onset (blue curve), a spectrum (black curve) was collected with changing pressure during the scan (which goes from right to left). The pressure initially started above onset, was adjusted below onset (at 1,940 nm), and then adjusted above onset (at 1,300 nm) during the scan. Asphaltene onset is reversible on a minutes time frame. From Joshi, Mullins, Jamaluddin, *et al.* (2001). © 2001 American Chemical Society.

For the third spectrum shown in Fig. 17 (black curve), the pressure was kept above onset initially, and when the spectral acquisition was at 1,940 nm (and moving to shorter wavelengths), the sample pressure was suddenly dropped from 13,000 psi to 6,000 psi, which is 2,000 psi below onset. Immediately, asphaltene flocculation resulted in substantial light scattering and indeed the resulting spectrum matched that previously acquired (with a different sample charge of the same live oil) at 6,000 psi. The spectral acquisition continued until 1,300 nm, when the sample pressure was suddenly increased to 13,000 psi, which is 5,000 psi above onset. Immediately, the scattering was reduced to zero, showing that the asphaltene flocculation is reversible on a minutes time frame (Joshi, Mullins, Jamaluddin, *et al.*, 2001), confirming other compelling work (Hammami, Phelps, Monger-McClure, *et al.*, 2000; Hammami and Ratulowski, 2007). Similar data of reversible light scattering with pressure changes would identify the source of scattering as asphaltenes. If the asphaltenes are allowed to settle to the bottom of the sample bottle and mix with potential clay and water to form an asphaltene mat, then no practical amount of sample agitation at high pressure can restore the crude oil to its initial behavior. The reversibility of asphaltene flocculation mandates redissolution on a minutes time frame. The same applies to field deposits of asphaltenes.

Crude oil color

Molecules that absorb light are called chromophores. A subset of chromophores fluoresce after absorbing light; these molecules are called fluorophores. This section discusses crude oil chromophores, which leads to the subsequent treatment of crude oil fluorophores.

Differentiation of crude oil from OBM filtrate downhole is accomplished in part by measuring color versus time. The OBM base oils that are commonly used span a wide range of different fluids, including diesel and mixtures of linear alpha olefins or internal olefins. (The olefin group disrupts molecular stacking, helping these high carbon number compounds to remain liquid—a rather important property of drilling fluids). Fortunately, typical formation crude oils contain two types of components that are not contained in almost any drilling fluid: asphaltenes and dissolved gas. Asphaltenes impart significant color to corresponding asphaltic materials—we are all familiar with the dark color of asphalt and roofing tar. The coloration of asphaltenes and similar colored compounds in crude oil derives from their electronic structure. A simplified description follows with the intent of giving a rough understanding of how color originates from electronic transitions in asphaltenes and the like.

Electrons, molecules, and other small particles are treated within the construct of quantum mechanics (Atkins and Friedman, 2005). This author attempts to provide a correct conceptual explanation of the facets of quantum mechanics that are relevant to crude oil coloration and DFA. Avoidance of the corresponding (and unlimited) complexity of quantum mechanics is assiduously attempted; consequently, it is accepted at the outset that this discussion is far from rigorous.

The intention is to impart heuristics to the reader regarding oil color based on molecular structure. Oil color is very important in all aspects of DFA, so this objective is merited. Electrons are described by wave functions; the square of the amplitude of the electron wave function in some spatial region gives the probability of the electron being in that volume.[8] An elementary quantum mechanical problem (perhaps an oxymoron) is to consider a "quantum particle in a box." The particle (for example, an electron) confined to a small "box," or basically spatially confined to a small volume, has a wave function of short wavelength. Short wavelengths correspond to high frequencies and high-frequency transitions. This is similar to guitar strings; as the nodes of the vibrating guitar string are moved closer, the wavelength of the guitar string resonance decreases and the acoustic frequency increases.

The electrons of the C–C and C–H bonds of saturated hydrocarbons are confined to a small volume, which localizes them in the immediate vicinity of the CC or CH atoms of the bond. This includes the CH_4 group, $-CH_3$ group, and $-CH_2-$ group, along with the C–C bond backbone of the alkanes. Within the sigma (σ) bonds of the C–C bonds (Fig. 18) the localized electrons have short wavelengths in their confined space and are thus characterized by high-frequency transitions. Consequently, saturated hydrocarbons are colorless. Examples include methane, pentane, and dodecane. Figure 18 is a representation of the σ-bonds of methane. Methane (and all alkane carbon) is tetrahedral, not square planar, in structure. Pure waxes are generally rich in long-chain n-alkanes, which are straight-chain alkanes with no branching groups. This geometric configuration is conducive to ordered molecular stacking such as a stack of logs, so the solids form at somewhat elevated temperatures and represent a wax problem in flow assurance. As noted previously, for OBM drilling fluids, the single olefin group of the linear alpha olefins disrupts this stacking, impeding wax formation. Pure waxes are transparent if melted or dissolved but white as a solid; there is no absorption of light that would impart color. The white appearance of the solid results from optical light scattering from the alternating amorphous and microcrystalline solid interfaces, with the ordered phase at slightly higher density. Oilfield waxes are typically dominated by n-alkanes but have the appearance of black shoe polish. This color originates from the small asphaltene content or entrained black oil in the waxes.

---

[8] *The treatment in quantum mechanics of finding electrons within a probabilistic, not deterministic, means was decried by Einstein in his famous quote "God does not play dice." Nevertheless, Einstein played a central role in establishing the foundations of quantum mechanics, for which he was awarded the Nobel Prize in Physics in 1921.*

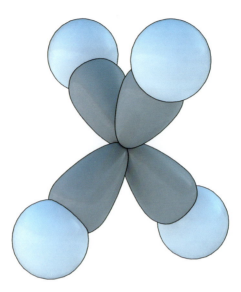

Figure 18. The electronic orbitals between the central carbon and four hydrogen atoms of methane have a tetrahedral arrangement. The electronic orbitals are localized, yielding high-energy transitions; consequently, methane and other alkanes are colorless. Chemical bonding originates from the constructive interference of the carbon and hydrogen electronic orbitals increasing electron density between the positive atomic cores.

In contrast to the localized electrons of the σ-bonds of saturated hydrocarbons, the electrons of the pi (π) bonds of aromatic compounds such as benzene are delocalized.[9] Also in contrast to the tetrahedral alkane carbon, all carbon and hydrogen atoms of benzene are in the same plane. As a consequence of delocalization, the π-electrons have much lower frequency transitions. Figure 19 is a representation of the π-electrons of benzene, which are in electron orbitals perpendicular to the molecular plane (the hydrogen atoms of benzene are not depicted). Compounds with these benzene rings are called aromatic compounds. There are also σ-bonds between the benzene carbon atoms (depicted as lines in Fig. 19) and between each carbon and its hydrogen (not depicted).

---

[9] *Chemical bonds are characterized by their symmetry. The σ-bonds are rotationally invariant around the bond axis, which is the C–H or C–C bond axis for hydrocarbons. The π-bonds between two atoms have a nodal plane (zero electron amplitude) containing the CH atoms and possess a mirror plane of symmetry perpendicular to this nodal plane.*

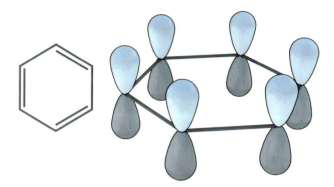

Figure 19. Benzene is the simplest aromatic compound. The six carbon atoms (shown) and six hydrogen atoms (not shown) reside in the same plane. The wave functions of the out-of-plane π-electrons are delocalized over all six carbon atoms of the benzene ring, which lowers the π-electron energies and energy transitions.

If two or more aromatic rings are fused, meaning that they share a common side, then the compound is called a polycyclic aromatic hydrocarbon (PAH). Figure 20 depicts naphthalene,[10] the simplest PAH.

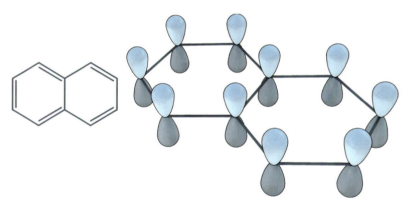

Figure 20. Naphthalene, with its two planar fused aromatic rings, is the simplest PAH. The π-electrons are delocalized over both rings; consequently, the lowest π-electronic transitions of naphthalene are lower energy than those of benzene.

---

[10] Naphthalene has been used in mothballs, the white balls placed with clothes in storage. Naphthalene sublimes (from the solid to gas), and the gas in sufficient concentration is deadly to moths. That naphthalene is carcinogenic and toxic to humans has reduced its popularity.

Moreover, the extent of delocalization is directly related to the number of rings in the aromatic ring system. The bigger the ring system, the greater the delocalization and the lower the frequency of the electronic transitions. Figure 21 shows the effect of the number of fused rings on the energy of electronic transitions of linear aromatic molecules, the acenes (Mullins, 1998). The electronic transition energy shifts to lower energy with the addition of each aromatic ring, and the larger ring systems have electronic transitions in the visible range. In this figure, fluorescence emission is plotted associated with the electronic transition from the first excited electronic state to the ground electronic state. The strong peaks in the spectrum of each compound correspond to the excitation of different vibrational states in the ground electronic state.[11]

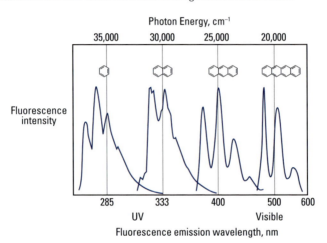

Figure 21. The lowest energy electronic transitions for a series of PAHs are shown. The larger the fused aromatic ring system, the longer the wavelength of the electron wave function and the lower the electronic transition in accord with the simple model of a quantum particle in a box. The larger compounds have electronic transitions in the visible range and are colored.

Single-ring aromatic compounds such as benzene are the simplest aromatic compounds in crude oil and have short-wavelength transitions (Fig. 21). Asphaltenes are the most complex and are the longest wavelength absorbers. The optical absorption spectrum of crude oils is produced by the contribution of all the aromatic compounds in crude oil. Figure 23 is an oft-reproduced example of crude oil coloration (courtesy of H. Elshahawi, Shell International E&P). Not only does the photograph illustrate the various crude oil colorations, but it is especially popular because all the oil samples originated from a single oil column.

---

[11] *The particular vibrational state of concern is the CC ring stretch mode. Changing the π-electron distribution by electronic excitation or de-excitation causes a change in the force constant of the ring stretch mode. Consequently, the ring stretch mode of the excited electronic state projects into many, not just one, of the vibrational wave functions of the ground electronic state and is described by the Franck-Condon factor (Atkins and Friedman, 2005). The energy spacings of the peaks for each compound in Fig. 21 correspond to the ring-stretch vibrational energies of the ground electronic state.*

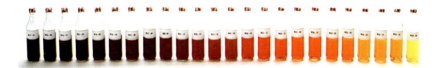

Figure 22. This photograph of the typical range of crude oil coloration is dramatic in that the crude oils all come from one oil column (see "The complexity of reservoir fluids" in the preceding chapter). Photograph courtesy of H. Elshahawi, Shell International E&P.

Figure 23 shows the optical (Vis-NIR) absorption spectra of very light to very heavy crude oils. In large measure, the crude oil colors in Fig. 22 correspond to the spectra in Fig. 23—or, more accurately, to the visible portion of these spectra (400 nm–700 nm)—even though different samples were used for the two figures. The spectra in Fig. 23 include much heavier and lighter oils than the photograph of Fig. 22.

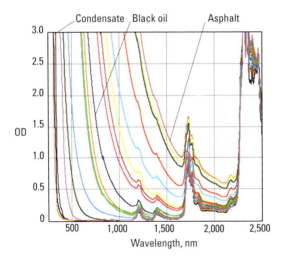

Figure 23. The Vis-NIR spectra of crude oils span the range from very light to very heavy. A huge variation in crude oil coloration is observed in accord with the huge variation in asphaltene and resin content. The NIR vibrational peaks are fairly similar for all (dead) crude oils. From Mullins, Mitra-Kirtley, and Zhu (1992).

As seen in Fig. 23, in the Vis-NIR range the crude oil spectra consist of a quantized series of vibrational bands, which are discussed in the next section. In addition, the crude oil spectra exhibit a highly variable, structureless, increasing absorption at shorter wavelengths. This absorption is due to the overlapping electronic transitions of the numerous PAHs in the oil. The electronic absorption edge, where appreciable electronic absorption first appears, is due primarily to asphaltenes and resins (the next heaviest component of crude oil). Because crude oils vary in heavy-ends content, the wavelength location of the electronic absorption edge for different crude oils is variable. For very light crude oils that do not contain any asphaltene or resin, the electronic edge is determined by the smaller aromatics.

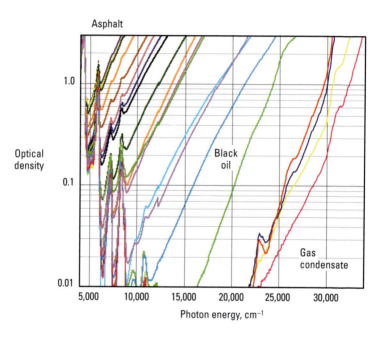

Figure 24. Replotting the data in Fig. 23 in a semilog plot versus photon energy shows that all crude oils have the same slope of their electronic absorption edge. This is reminiscent of the Urbach tail in solid-state physics and gives the population distribution of large chromophores. From Mullins, Mitra-Kirtley, and Zhu (1992).

If the data in Fig. 23 is replotted in a semilog plot versus photon energy, then all crude oils, from light condensates to heavy tars, exhibit a straight line of the same slope (Fig. 24) (Mullins, Mitra-Kirtley, and Zhu, 1992). In addition, asphaltenes exhibit this same phenomenon (Mullins and Zhu, 1992). This remarkable result is reminiscent of the Urbach tail from solid-state physics (Urbach, 1953). The Urbach equation for the electronic absorption edge is

$$\alpha = \alpha_o \exp\left(\frac{h\nu}{E_u}\right), \tag{10}$$

where α is the OD at photon frequency ν (see Eq. 3), α_o is a reference optical density, E_u is the Urbach decay constant, and h is Planck's constant.

A semiconductor at a temperature of absolute zero has an infinitely sharp Fermi edge (the electronic absorption edge). At photon energies slightly less than the electronic excitation energy (band-gap), there is no optical absorption. At photon energies at and above the electronic excitation energy, there is large absorption. At finite temperatures, the Fermi edge exhibits the Urbach tail, which represents a Boltzman distribution of the thermal excitation of absorber sites (Urbach, 1953). That is, the thermal excitation of any absorber site in the semiconductor has its excitation energy reduced by just that thermal excitation. Plotting the log of OD versus photon energy for the Urbach region gives a slope of kT; that is, $E_u = kT$ in Eq. 10.

The result that crude oil spectra obey principles from solid-state physics is surprising. For crude oils and asphaltenes, the Urbach slope is $E_u \sim 10kT$ (for room temperature measurements). This is not surprising—no one is suggesting that crude oils are black because of the presence of very hot benzene. What is the connection of the Urbach tail with crude oil? Oil results from the catagenesis of kerogen (Tissot and Welte, 1984), the insoluble organic matter of sedimentary rock. Catagenesis is primarily driven by heat over geologic time. The generation of crude oil chromophores corresponds to the thermal production of larger PAHs from smaller ones (Mullins, Mitra-Kirtley, and Zhu, 1992). And, of course, the larger PAHs absorb at longer wavelengths in accord with the quantum particle in a box model (Atkins and Friedman, 2005). This process relates to the colors that occur in toasting white bread. As toasting proceeds, the color evolves— white, yellow, tan, brown, and then black, if the toaster is on too long.

In particular, the electronic absorption edge of crude oils and of asphaltenes is given by the Urbach formalism. Analysis of the relationship of the optical absorption and fluorescence emission processes in the Urbach region has established that the Urbach tail in crude oils is associated with an exponential decrease of longer wavelength absorbers. As simplified for pedagogic purposes in Fig. 25, there is an exponential increase in color-coded chromophore population—thus an Urbach spectral region—until benzene is encountered. The population of benzene is depicted as roughly equal to that of naphthalene. The Vis-NIR electronic absorption spectrum of crude oils and asphaltenes is composed of the sum of spectra of the constituent aromatic compounds. Crude oil chromophores are discussed in the next section.

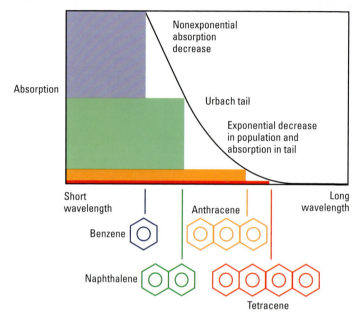

Figure 25. As shown in a simplified representation, the Urbach tail in crude oils is associated with the exponential reduction of optical absorption at (color-coded) long wavelengths resulting from a corresponding exponential decrease of the longest wavelength absorber molecules. Larger, redder chromophores are thermally produced in exponentially declining numbers from smaller, bluer chromophores in crude oils and asphaltenes. The non-Urbach behavior of the short-wavelength region occurs because the corresponding small chromophores are not exponentially more populous (Mullins, 1998). With kind permission of Springer Science+Business Media.

Figure 26 shows band locations of PAHs that are likely to be found in crude oils and asphaltenes. Indeed, Fig. 26 shows that the absorption peaks are dependent on the number of fused aromatic rings in the PAH (Ruiz-Morales and Mullins, 2007). Another more subtle point is that the location of the PAH absorption peaks also depends on ring system geometry (Ruiz-Morales, 2007). Detailed analysis of crude oil chromophores by X-ray Raman spectroscopy shows that they are pericondensed (in essence, more circular PAHs), not catacondensed (more linear PAHs, similar to the acenes) (Bergmann, Groenzin, Mullins, *et al.*, 2003). Detailed molecular orbital calculations show that this follows from energy stability arguments (Ruiz-Morales, 2007). The PAHs in Fig. 26 are energetically stable and thus can survive over geologic time. Several examples of the PAH ring systems are also shown. It is these ring systems that are present in decreasing concentration with increasing ring size, thereby accounting for the Urbach formalism (Mullins, 1998).

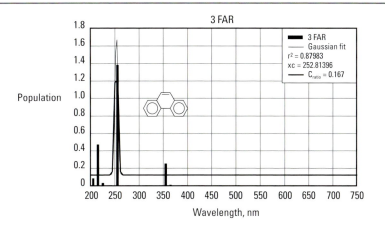

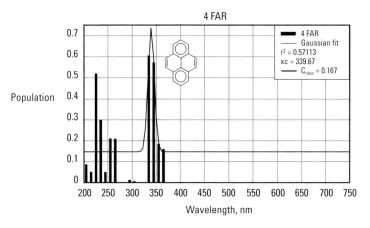

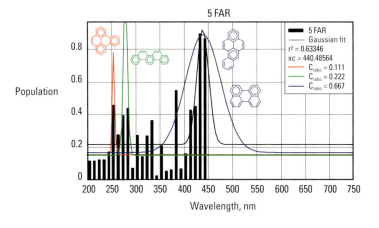

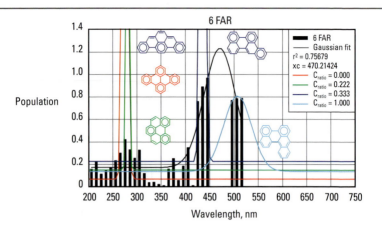
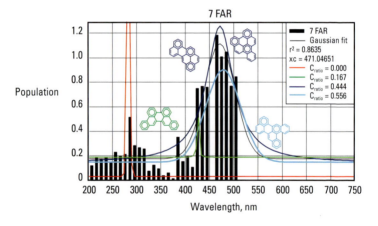
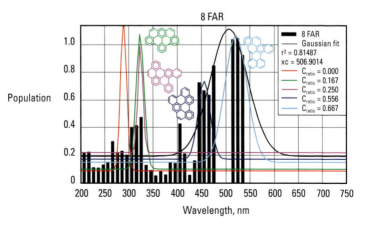

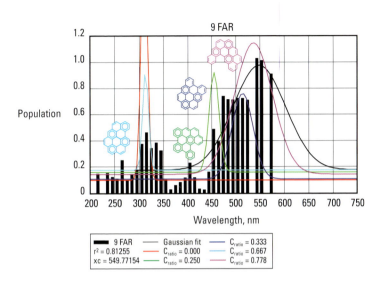

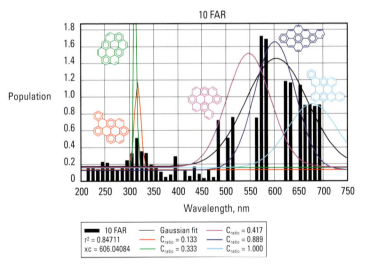

Figure 26. Absorption band spectral location is a function of the number of fused aromatic rings (FAR) in PAHs. Larger ring systems generally have redder transitions in accord with the quantum particle in a box model. The pericondensed PAH ring systems (more circular ring geometry) are energetically stable and are good candidates for crude oil and asphaltene PAHs. From Ruiz-Morales and Mullins (2007). © 2007 American Chemical Society.

The largest PAHs in crude oil are contained in asphaltene molecules. Various measurements show that most asphaltene chromophores possess 4 to 10 fused aromatic rings, with 7 fused rings being the most probable. Although this result has been controversial, it is supported by many studies, including time-resolved fluorescence depolarization (Groenzin and Mullins, 1999, 2000), scanning tunneling microscopy molecular imaging (Zajac, Sethi, Joseph, *et al.*, 1994), high-resolution transmission electron microscopy molecular imaging (Sharma, Groenzin, Tomita, *et al.*, 2002), molecular orbital theory with optical spectra (Ruiz-Morales and Mullins, 2007), and fragmentation mass spectroscopy (Rogers, Tan, Ehrmann, *et al.*, 2008). Candidate asphaltene molecular structures are in Fig. 27 (Groenzin and Mullins, 2000). Results from fluorescence correlation spectroscopy of asphaltenes corroborate these results (Andrews, Guerra, Mullins, *et al.*, 2006).

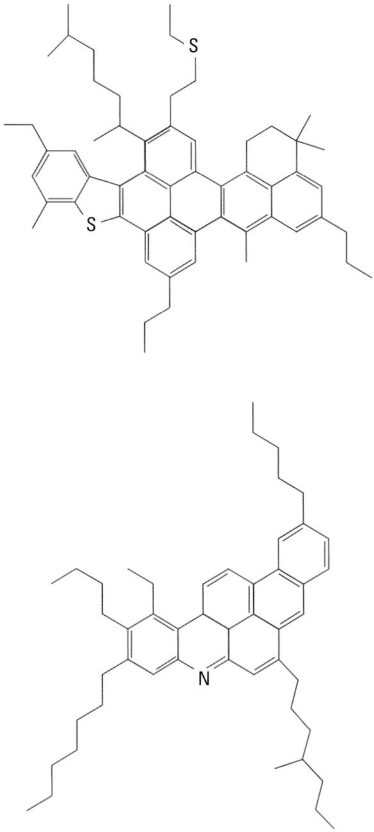

Figure 27. Although asphaltene samples contain at least tens of thousands of unique compounds, the bulk of asphaltene molecules share traits with these candidate molecular structures. From Groenzin and Mullins (2000). © 2000 American Chemical Society.

Temperature dependence of crude oil coloration

The temperature dependence of the electronic absorption of PAHs is almost nil (Fig. 28). The Boltzmann distribution is one of the foundations of statistical mechanics and gives the population distribution P_i for states i of energy ΔE above the ground state. $P_i = 1$ for the ground state ($\Delta E = 0$). For equal numbers of states at energy E, the population is given by the Boltzmann factor:

$$P_i = \exp\left[-\frac{\Delta E}{kT}\right], \qquad (11)$$

where k is Boltzmann's constant (~0.7 cm^{-1}/K) and T is temperature. Typical electronic transitions are ~20,000 cm^{-1}; thus, all electronic population is in the ground state.

Figure 28 shows the small and reversible temperature effect for crude oil coloration (Ruiz, Wu, and Mullins, 2007). The largest change is associated with thermal expansion. The thermal expansion coefficient for dead crude oils is ~10^{-3}/degC; thus, there is on the order of a 10% reduction of crude oil color compared at room temperature (the spectra at elevated temperature were obtained at modest applied pressure). The thermal expansion coefficient was not explicitly measured for these crude oils.[12] This and the concentration data that follow help rule out any significant contribution of charge transfer to crude oil coloration (Ruiz, Wu, and Mullins, 2007).

In addition, asphaltene and crude oil coloration is trivially dependent on concentration. For example, Fig. 29 shows that the measured color of an asphaltene solution equals the color measured for a solution diluted by 100× but with a 100× increase in pathlength. That is, crude oil coloration obeys the Beer-Lambert law (Eq. 8). In the top panel of Fig. 29, the two spectra overlie where the concentration 50 mg/100 mL exceeds the critical nanoaggregate concentration (CNAC), whereas the 100× dilution is less than the CNAC. Thus, asphaltene aggregate formation does not affect the color measurement. The bottom panel shows that at much higher concentrations, there is still no nonlinear color dependence on concentration. The linearity of asphaltene color allows using color measurement to determine relative asphaltene concentration in crude oils, in turn enabling DFA methods to track changes in asphaltene content, for instance, from gravity.

---

[12] *The compressibility for dead crude oils is ~10^{-4}/atm and the thermal expansion is ~10^{-3}/degC. Thus, the counteracting density change of pressure and temperature is ~150 psi/degC. For a fixed volume (and no phase change), a temperature reduction of 10 degC gives a 1,500-psi pressure reduction. Thus, samples stored in high-pressure bottles of fixed volume often undergo phase changes when cooled. Live crude oils have higher compressibilities, so they have correspondingly smaller pressure changes.*

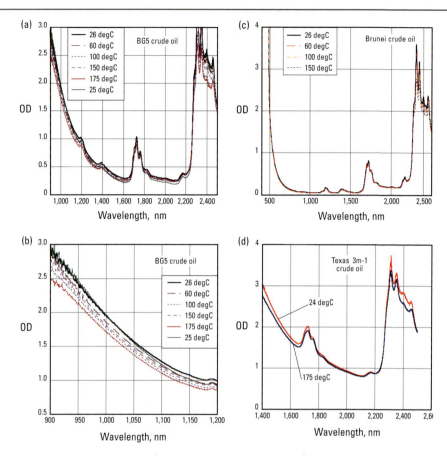

Figure 28. The predominant temperature dependence of the coloration of light, medium, and heavy crude oils is the trivial and reversible thermal expansion effect of the bulk sample. Coloration specifically dependent on molecular complex formation is ruled out. From Ruiz, Wu, and Mullins (2007). © 2007 American Chemical Society.

Quicksilver Probe focused fluid acquisition and asphaltenes

The fact that crude oil coloration depends trivially on concentration is the basis of the OCM algorithm for contamination monitoring. The dilution of crude oil with OBM filtrate reduces the resulting coloration linearly with the dilution. However, very large concentrations of OBM filtrate could destabilize the asphaltenes, resulting in their precipitation. Indeed, the standard laboratory procedure for isolating asphaltenes is precipitation with n-alkanes, which are chemically highly similar to OBM filtrate. In this precipitation case, the color is not linearly dependent on dilution. In addition, the collected sample, having lost asphaltenes, is not representative. This situation cautions that the use of Quicksilver Probe focused fluid acquisition is desirable in order to sample reservoir fluids that have not been in contact with OBM filtrate.

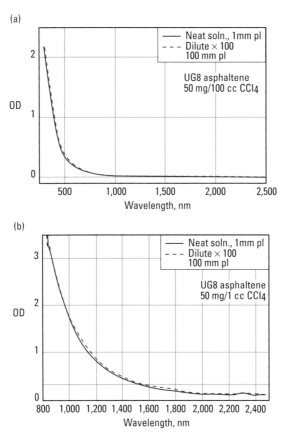

Figure 29. Asphaltene color depends linearly on concentration. The top panel compares spectra for the concentration 50 mg/100 mL and a 1-mm pathlength with 0.5 mg/100 mL at a 100-mm pathlength. The bottom panel compares 50 mg/1 mL at a 1-mm pathlength with 0.5 mg/1 mL at a 100-mm pathlength. From Ruiz-Morales, Wu, and Mullins (2007). © 2007 American Chemical Society.

Hydrocarbon compositional analysis

Vibrations of molecules can be thought of as mechanical oscillations of masses and springs, with the masses being atoms and the springs being their chemical bonds. Masses connected by springs obey Hooke's law:

$$F = -kx, \qquad (12)$$

where F is force, k is the spring force constant, and x is the length displacement from its resting or equilibrium length. (Compression and expansion of the spring are presumed.) The negative sign means that the force is a restoring force. The potential energy E_p of the spring is given by

$$E_p = \frac{1}{2}kx^2. \qquad (13)$$

The force (Eq. 12) is given by the negative gradient of the potential (Eq. 13). The corresponding potential energy curve versus equilibrium length is simply quadratic (Fig. 30), symmetric about $x = 0$, and called the "harmonic oscillator." Figure 30 also shows the potential energy curve for a molecular bond; at low energies the molecular bond is close to the harmonic oscillator. At high energies, the chemical bond can dissociate (the spring breaks), and once this happens there is no more restoring force (the outer turning point in Fig. 30). At the inner turning point at high energies, molecules exhibit more repulsion than the harmonic oscillator.

Figure 30 shows that vibrational levels of molecules are quantized. Similar to electrons, molecular vibrations are also described by wave functions (Atkins and Freidman, 2005). Here, quantization results because half-integral vibrational wavelengths (for example, ½, 1, 1½) can fit in the potential at increasing energies, but other wavelengths (for example, 1.3) do not fit. There must always be negligible wave function amplitude at the turning points. (For classic oscillators, their differing vibrational energies overlap, resulting in a continuum of possible energies.)

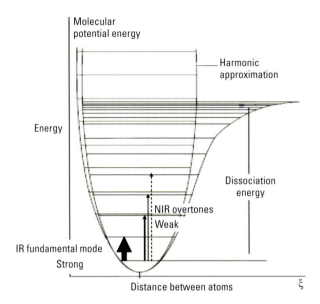

Figure 30. The potential energy curves of the harmonic oscillator and the chemical bond are shown versus bond length ξ. The molecular potential resembles the harmonic oscillator at lower energies. The only allowed radiative transitions are between adjacent (quantized) vibrational states. For reasonably energetic vibrations, only the ground vibrational state is appreciably occupied at room and borehole temperatures.

The vibrational frequency ν is given by

$$\nu = \sqrt{\frac{k_F}{\mu}}, \tag{14}$$

where k_F is the vibrational bond force constant and μ is the reduced mass. Similar expressions apply to any mechanical oscillator.[13] Fundamental vibrational modes of organic molecules occur in the MIR spectral range. That is, for organic molecules, the fundamental energies of vibrations

---

[13] A similar expression applies for acoustic frequencies. The speed of sound u is given by $u = \sqrt{1/\beta\rho}$, where β is compressibility (or the inverse of the bulk modulus) and ρ is density. The acoustic frequency is u/λ, where λ is the wavelength (see Eq. 2).

are generally in the 400 cm^{-1} to 4,000 cm^{-1} range. And the MIR range is useful for chemical group identification because of the differing absorption frequencies of the different chemical groups. Low overtones of the highest energy transitions occur in the NIR, and only low overtones carry enough amplitude to be easily seen (Fig. 30). Bonds involving hydrogen atoms have reasonably large force constants and small reduced mass, giving these bonds the largest frequencies (see Eq. 14). Correspondingly, low-overtone C–H bonds occur in the NIR; these overtones also carry chemical group identity like the MIR bands, in particular, CH_4, $-CH_3$, and $-CH_2-$ compositional information.

The photoexcitation of vibrations—so important to DFA—is out of the ground vibrational state of the high-energy CH oscillators (Fig. 30). Temperature dependence is addressed in greater detail subsequently. A fundamental property of harmonic oscillator wave functions is that photoexcitation can cause transitions only between adjacent levels. Thus, from the ground state, the only allowed photoexcitation process is to the first excited state, which corresponds to the MIR range for crude oils. Because the actual molecular potential deviates slightly from the harmonic oscillator,[14] nominally forbidden transitions occur but are quite weak. Their absorption coefficient is small, so long pathlengths are needed to compensate in order to have appreciable absorption (see Eqs. 7 and 8). Long pathlengths (millimeter scale) are desirable for DFA and other process streams.

The low overtones pose a complication. The fundamental vibrational modes are described accurately by normal mode analysis, giving simple Lorentzian or Voigt peaks. High overtones are well described by local mode analysis and again simple peaks are obtained. However, low overtones are not well described by either formalism, and complex band shapes are obtained. Nevertheless, standard chemometrics can be applied, and the NIR spectral region has proved valuable for many types of chemical analysis (Malinowski, 1991), particularly on process flow streams. For example, the Btu content of natural gas (Brown and Lo, 1993) and the aromatic content of gasoline (Cooper, Wise, Welch, *et al.*, 1997) have been correlated to the NIR spectrum.

Figure 31 shows that the Vis-NIR spectra of *n*-heptane and of water are distinct. For heptane, both the $-CH_2-$ and $-CH_3$ chemical groups contribute to the spectra. The NIR spectra are characterized by a set of decreasing peaks at shorter wavelengths, with the spacing exhibiting the quantized nature of vibrational transitions, as discussed previously. Because these vibrational transitions are formally forbidden, they are weak. Higher overtones are more forbidden (Fig. 31), and thus the peaks are smaller at higher energies. For heptane, the band identities are the CH stretch+bend at 2,300 nm, two-stretch at 1,725 nm, two-stretch+bend at 1,400 nm, and three-stretch at 1,190 nm. Water exhibits the OH stretch+bend at 1,900 nm and two-stretch at 1,450 nm. The two-stretch peaks of oil and water are of convenient magnitude and location for flowline analysis.

---

[14] *Forbidden transitions also occur as a result of the nonlinear change in molecular dipole moment with the change in bond distance.*

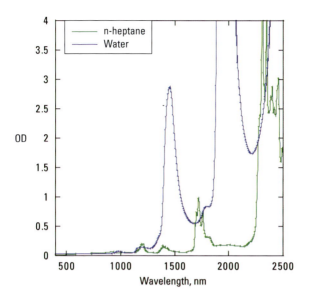

Figure 31. The Vis-NIR spectra of *n*-heptane and water for a 2-mm pathlength are distinct. The CH overtone vibrational band at 1,725 nm is particularly useful for DFA measurement applications (Safinya and Tarvin, 1991).

Figure 32 shows the CH two-stretch overtone spectral region of methane, *n*-heptane, and a mixture of the methane and *n*-heptane (Mullins, Daigle, Crowell, *et al.*, 2001). The analysis of many straight-chain and branched hydrocarbons enabled determination that the two high-energy peaks at 1,690 nm and 1,700 nm are due to the $-CH_3$ moiety, whereas the large peak at 1,721 nm is predominantly $-CH_2-$ in character (Fujisawa and Mullins, 2007; O.C. Mullins, pers. comm., 1989). Thus, the NIR peaks for the three chemical groups CH_4, $-CH_3$, and $-CH_2-$ are sufficiently spectrally resolved and in order in accord with their reduced mass, enabling conceptually simple chemical analysis. Detailed analysis of Fig. 32 shows that the spectrum of the mixture of methane and *n*-heptane equals the weighted sum of the spectra of the components (Mullins, Joshi, Groenzin, *et al.*, 2000). Spectral addition applies because the hydrocarbons interact so weakly. Indeed, alkanes are viewed as ideal fluids because they are so weakly interacting. This spectral linearity breaks down at low pressures (1,000 psi), where the methane spectrum starts to exhibit individual rotational-vibrational spectral lines (O.C. Mullins, pers. comm., 1990). However, most oil and gas reservoirs are at elevated pressures, mitigating the pressure dependence of the methane spectra at low pressure. No other hydrocarbon exhibits this problem at pressures of interest.

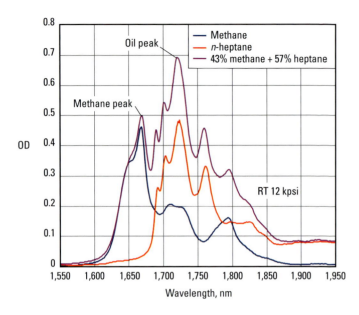

Figure 32. In the NIR CH two-stretch overtone, the methane peak is distinguished from peaks of other alkanes. Spectral linearity for alkanes is observed. Detailed analysis of this CH two-stretch overtone yields compositional information of crude oils. From Mullins, Daigle, Crowell, *et al.* (2001).

One of the most important attributes of a crude oil is the GOR. The GOR of a single-phase hydrocarbon fluid from the reservoir is defined as the volumetric ratio of gas to liquid oil at conditions of 1 atm and 60 degF. The units of GOR are standard cubic feet of gas per barrel of liquid, or meter cubed per meter cubed. Figure 33 shows the spectra of several live crude oils of differing GOR (Mullins, Daigle, Crowell, *et al.*, 2001). The spectra of methane and a dead crude oil are also plotted. It has been shown that the GOR of crude oils can be determined by NIR spectral analysis (Mullins, Daigle, Crowell, *et al.*, 2001; Mullins, Beck, Cribbs, *et al.*, 2001; Dong, Mullins, Hegeman, *et al.*, 2002). These concepts have been applied in the oilfield (Mullins, Daigle, Crowell, *et al.*, 2001), where recent refinements to quantitative spectral analysis have proved particularly effective and are in use worldwide in oil wells with the LFA Live Fluid Analyzer module of the MDT toolstring (Dong, Mullins, Hegeman, *et al.*, 2002). The CFA Composition Fluid Analyzer extended the ability to measure GOR to higher GOR fluids. More recently, the IFA* Insitu Fluid Analyzer has been introduced to the field, with many additional spectral channels that greatly improve accuracy, repeatability, and resolution (Dong, O'Keefe, Elshahawi, *et al.*, 2007). Specifically, the IFA analyzer facilitates comparison of data from different wells from different times, making the IFA analyzer the focus of DFA applied to reservoirs, not just single wells.

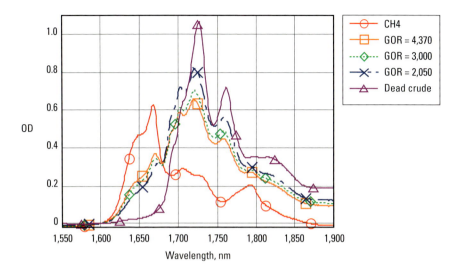

Figure 33. The NIR CH overtone differentiates several live crude oils (with dissolved gases) of different GORs along with the spectra of methane and a dead crude oil (with no dissolved gas). From Mullins, Daigle, Crowell, et al. (2001).

The simplest interpretation algorithm for obtaining GOR uses just two optical channels (plus a baseline channel), which is convenient for oilfield applications. GOR can be obtained from first principles (Fig. 34), as subsequently described. GOR was introduced to the industry as the first true DFA measurement (Mullins, Daigle, Crowell, et al., 2001; Mullins, Beck, Cribbs, et al., 2001; Dong, Mullins, Hegeman, et al., 2002). The downhole Vis-NIR measurements that preceded the GOR measurement performed fluid identification measurements designed primarily for the identification of hydrocarbon, determination of liquid or gas, and acquisition of valid samples. To obtain GOR, one channel can be placed on the methane peak at ~1,670 nm and another channel can be placed on the "oil" peak at ~1,725 nm, which is the overtone with predominant –CH_2– character. A third channel is used for the baseline measurement. Naturally, the use of more channels can improve the measurement (Mullins, Joshi, Groenzin, et al., 2000).

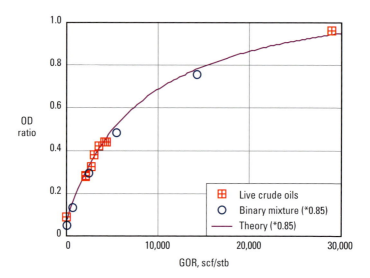

Figure 34. Quantitative analysis of the NIR overtone of single-phase live crude oils enables the determination of their GOR values. From Mullins, Daigle, Crowell, *et al.* (2001).

To calculate GOR, define a compositional mass vector (Mullins, Daigle, Crowell, *et al.*, 2001; Mullins, Beck, Cribbs, *et al.*, 2001; Dong, Mullins, Hegeman, *et al.*, 2002; Fujisawa, van Agthoven, Jenet, *et al.*, 2002):

$$\mathbf{m} = \begin{pmatrix} m_m \\ m_o \end{pmatrix} \qquad (15)$$

and spectral signal vector

$$\mathbf{S} = \begin{pmatrix} s_m \\ s_o \end{pmatrix}. \qquad (16)$$

A 2×2 spectral response matrix **B** relates **m** and **S**:

$$\mathbf{B} = \begin{pmatrix} b_{mm} & b_{mo} \\ b_{om} & b_{oo} \end{pmatrix}, \qquad (17)$$

where b_{mm} and b_{mo} are the response factors of methane channel to the masses of methane and oil, respectively, and b_{om} and b_{oo} are the response factors of oil channel to the masses of methane and oil, respectively.

From the measurements of more than 30 oils and condensates with the LFA tool, the **B** matrix has been resolved. GOR can be calculated from the mass fractions of methane and oil by (Dong, Mullins, Hegeman, *et al.*, 2002)

$$GOR = 8{,}930 \; \frac{\dfrac{m_m}{m_o}}{1 - 0.193 \dfrac{m_m}{m_o}} \; \text{scf/bbl.} \tag{18}$$

There are no pressure or temperature terms in this GOR equation, making field interpretation relatively simple. GOR is now routinely measured on crude oils in the wellbore by using the LFA tool on the MDT platform. For crude oils with a high degree of coloration, the coloration can be measured and its impact subtracted by using the Urbach equation (Eq. 10) from the crude oil spectrum to calculate GOR (Dong, Mullins, Hegeman, *et al.*, 2002). Likewise, the presence of small amounts of water can also affect the spectrum in predictable ways. This effect too can be subtracted, enabling accurate GOR determination (Dong, Mullins, Hegeman, *et al.*, 2002).

Further compositional analysis can be accomplished by using more optical channels. Figure 32 shows that the two-stretch overtones of the CH_4, $-CH_3$, and $-CH_2-$ groups differ. Because there are three predominant alkane signatures, the expectation is that three hydrocarbon components (or pseudocomponents) can be extracted. In particular, CH_4 is methane, $-CH_3$ groups are highly enriched in light alkane gases, and $-CH_2-$ is predominant in hydrocarbon liquids (Fujisawa, van Agthoven, Jenet, *et al.*, 2002). Detailed spectral analysis can yield even more hydrocarbon compositional determination. In particular, ethane has unique spectral features that enable its analysis. In addition, CO_2 can be determined by using its NIR signature (van Agthoven, Fujisawa, Rabbito, *et al.*, 2002) (Fig. 35).

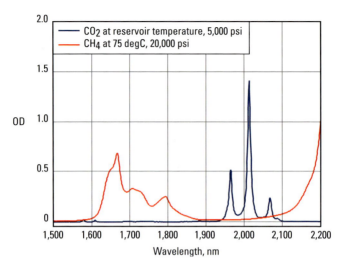

Figure 35. The NIR spectra of CH_4 and CO_2 are quite distinct, enabling their analysis (van Agthoven, Fujisawa, Rabbito, et al., 2002).

Principal components analysis (PCA) and principal components regression (PCR) can be employed for the data analysis and model building. Other chemometric techniques such as partial least square (PLS) should give similar results for this type of problem. These analytical techniques are explained well in many standard chemometrics textbooks and the related literature (Malinowski, 1991). For the purposes of the CFA Composition Fluid Analyzer module of the MDT tool, the hydrocarbon fluid is analyzed in terms of concentrations for four components or pseudocomponents: methane (C_1), nonmethane hydrocarbon gases (C_2–C_5), liquids (C_{6+}), and carbon dioxide (CO_2) (Fujisawa, van Agthoven, Jenet, et al., 2002). The oil coloration measured by the LFA analyzer is proportional to the asphaltene content, but that is analyzed separately. The CFA analysis algorithm was constructed in the following manner (Fujisawa and Mullins, 2007). First, NIR spectra for 20 different, known hydrocarbon samples were measured at pressures ranging from 4,000 psi to 15,000 psi at temperatures ranging from 20 degC to 150 degC. This data was used to construct the calibration data matrix $D_{t \times m}$, where t represents the number of channels and m represents the number of the calibration spectra.[15] Following the standard PCA procedure, the $D_{t \times m}$ matrix is decomposed into two orthogonal matrices, $R_{t \times t}$ (loadings) and $C_{t \times m}$ (scores):

$$D_{t \times m} = R_{t \times t} C_{t \times m}. \tag{19}$$

[15] Per convention, a matrix is denoted with a bold capital letter followed by two suffix numbers showing the size of the matrix. A vector is expressed with a bold lowercase letter, and a scalar is expressed with an italic letter. For the sake of consistency, suffix numbers are also added to vectors and scalars.

There are more samples than channels ($t < m$), which is unusual for spectrometers in laboratories but is the case for wireline tool spectrometers owing to the limited channel numbers. Among t principal components, only f of them are useful for predicting the concentrations of composition groups. So the dimensions of $\mathbf{R}_{t \times t}$ and $\mathbf{C}_{t \times m}$ can be reduced without losing meaningful information. Now the data matrix $\mathbf{D}_{t \times m}$ is approximated by $\underline{\mathbf{D}}_{t \times m}$:

$$\mathbf{D}_{t \times m} \sim \underline{\mathbf{D}}_{t \times m} = \underline{\mathbf{R}}_{t \times f} \underline{\mathbf{C}}_{f \times m}, \tag{20}$$

where $\underline{\mathbf{R}}_{t \times f}$ and $\underline{\mathbf{C}}_{f \times m}$ represent the reduced loading matrix and the reduced score matrix, respectively. In the PCR model, the concentration of each chemical composition $\mathbf{y}$ is independently related to the score of spectra by a regression vector $\mathbf{b}$. Using the m calibration dataset, the relation is

$$\mathbf{y}_{1 \times m} = \mathbf{b}_{1 \times f} \underline{\mathbf{C}}_{f \times m}. \tag{21}$$

Because both $\mathbf{y}_{1 \times m}$ and $\underline{\mathbf{C}}_{f \times m}$ are known quantities, the regression vector $\mathbf{b}_{1 \times f}$ can be determined using a least-squares approach. Once the regression vector $\mathbf{b}_{1 \times f}$ is determined from the calibration dataset, the concentration of each composition group in an unknown reservoir fluid is predicted by the following model equation:

$$\mathbf{y}_{1 \times 1} = \mathbf{b}_{1 \times f} (\underline{\mathbf{R}}_{t \times f}^T \underline{\mathbf{R}}_{t \times f})^{-1} \underline{\mathbf{R}}_{t \times f}^T \mathbf{m}_{t \times 1}, \tag{22}$$

where $\mathbf{m}_{t \times 1}$ is a spectrum of the reservoir fluid measured with a t-channel wireline spectrometer tool.

The CFA tool employs these measurements, and the corresponding tool analyses based on this model agree well with laboratory analysis (Fig. 36). In addition, the CFA tool performs these compositional measurements successfully in the field (Fujisawa, Mullins, Dong, et al., 2003; Fujisawa, Betancourt, Mullins, et al., 2004). The CFA data gives a more accurate fluid GOR for high-GOR fluids such as retrograde gas. For relatively low-GOR fluids, both the LFA and CFA tools naturally give similar analysis results because the influence of the C_2–C_5 group concentration to the overall fluid GOR is very small for those fluids.

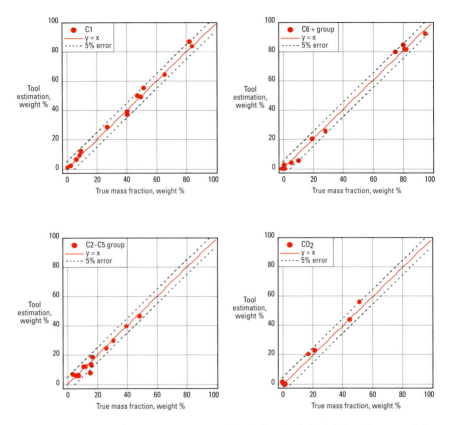

Figure 36. The results of CFA analysis compare well for C_1 (methane), C_2–C_5 (other alkane gases), C_{6+} (hydrocarbon liquids), and CO_2 for many different crude oils and prepared mixtures. From Fujisawa, van Agthoven, Jenet, *et al.* (2002).

The primary spectral effect of pressure on hydrocarbons is to change the mass density (Fig. 37). Likewise, the primary spectral effect of temperature is on mass density (Fig. 38). For hydrocarbons at room temperature, $kT \sim 200$ cm^{-1}, and for the temperatures found in oil wells, $kT < 400$ cm^{-1}. This energy is small compared with C–H vibrational energies of ~3,000 cm^{-1}. Correspondingly, the Boltzmann distribution (Eq. 11), shows that the bulk of the molecular population of C–H oscillators is predominantly in the ground vibrational state. This helps reduce any temperature dependence of the NIR spectra of hydrocarbons (Fig. 38).

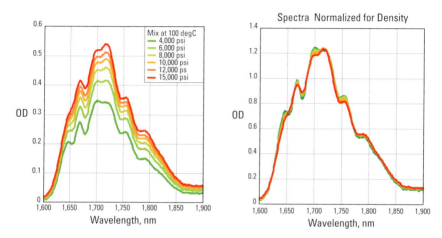

Figure 37. The primary effect of pressure at fixed temperature on the NIR spectra of a hydrocarbon mixture is to change the mass density. The spectra at different pressures nearly overlie when they are mass normalized (Fujisawa, van Agthoven, Jenet, *et al.,* 2002).

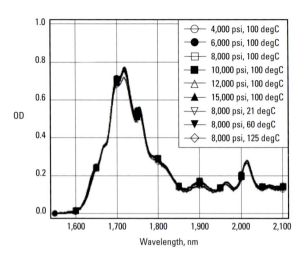

Figure 38. Normalized by mass density, the NIR spectra of a hydrocarbon mixture at various temperatures and pressures are nearly invariant. From Fujisawa, van Agthoven, Jenet, *et al.* (2002).

Finally, water has an interesting absorption spectrum. Figure 39 shows the absorption spectrum of water over a broad spectral range (Jackson, 1999). The somewhat narrow transmission band of water is where the eye has sensitivity. Of course, the eye contains water and evolved in water. Moreover, the transmission band of water coincides with the maximum output of the sun.

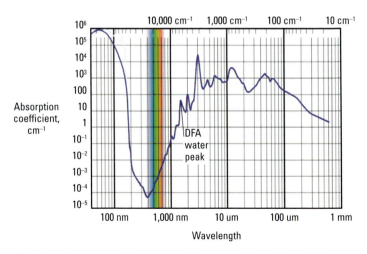

Figure 39. The absorption spectrum of water spans a huge range of absorption coefficients over a broad spectral range (Jackson, 1999). Water is opaque at most frequencies outside of the visible.

Index of refraction and gas detection

Hydrocarbon phase transitions: Gas

Figure 40 is a phase diagram of a typical formation crude oil with the critical point indicated where, by definition, the gas and liquid phases have identical properties (McCain, 1990). Inside the lobe is the two-phase region. Lines of constant gas and liquid mass fractions are indicated on the phase diagram; all these lines converge onto the critical point. To the left of the critical point is the bubblepoint line, which is the 100% liquid line. Consider a formation hydrocarbon that exists in the formation at the temperature and pressure at point A in Fig. 40. To flow oil out of the formation, a pressure drop is required, indicated by the arrow pointing down from A in the figure. If the pressure drop is too large, then the bubblepoint line is intersected, producing a second hydrocarbon phase: gas. Generally, gas bubbles form in a continuous liquid phase. It is very undesirable to have such gas evolution occur in the formation. Because gas is lower viscosity than oil, it flows preferentially after exceeding the critical gas saturation, leaving the valuable liquid hydrocarbon in the formation. In any event, if the sampling process yields a second phase, it is highly unlikely that the corresponding sample is representative. Thus, it is desirable to sample and eventually produce the formation above the bubblepoint.

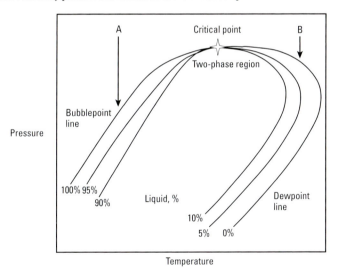

Figure 40. This prototypical phase diagram of a crude oil shows the bubblepoint line (100% liquid) and the dewpoint line (0% liquid) meeting at the critical point. The generation of a second phase in sample acquisition is to be avoided because then the sample is not representative of the formation fluids.

Gas can be differentiated from liquid by index-of-refraction measurements. Light incident on an interface between two nonabsorptive materials (dielectrics) undergoes reflection and refraction. A smooth, flat surface produces specular reflection, where the angle of incidence equals the angle of reflection (Fig. 41). The angle of refraction is given by Snell's law, where n_1 and n_2 are the indices of refraction:

$$n_1 \sin(\theta_1) = n_2 \sin(\theta_2). \tag{23}$$

Light has two polarizations, and these can be projected into two perpendicular linear polarizations or two oppositely rotating circular polarizations. Linear polarization is used for the gas detector, in particular, p-polarization, in which the electric vector is in the plane of incidence and reflection. The amplitude of reflected p-polarized light is (Jenkins and White, 2001)

$$\sqrt{\mathfrak{R}} = \frac{\left(\dfrac{n_1}{n_2}\right)^2 \cos(\theta_i) - \left[\left(\dfrac{n_1}{n_2}\right)^2 - \sin^2(\theta_i)\right]^{1/2}}{\left(\dfrac{n_1}{n_2}\right)^2 \cos(\theta_i) + \left[\left(\dfrac{n_1}{n_2}\right)^2 - \sin^2(\theta_i)\right]^{1/2}}. \tag{24}$$

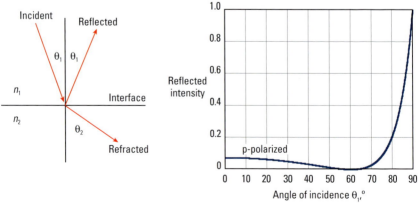

Figure 41. For the geometry pertinent to gas detection, the reflection intensity for p-polarized light is shown versus the angle of incidence. The reflected intensity goes through a minimum of zero and maximum of one over a small angular range.

For $n_1 < n_2$, all light is reflected at 90°. For $n_1 > n_2$, the light in Fig. 41 originates in the higher index medium and all light is reflected for all angles greater than the critical angle θ_c (Fig. 42). Called "total internal reflection," this principle is employed in fiber optics, for example, to obtain very small losses over very long distances. The value of the critical angle is

$$\theta_c = \sin^{-1}\left(\frac{n_2}{n_1}\right). \quad (25)$$

For sapphire windows, which are used in all DFA optics tools, $n_1 \sim 1.74$ (slightly wavelength dependent) and the critical angle for low-pressure gas ($n_2 \sim 1$) is 35°. The reflection curve for p-polarized light in Fig. 41 shows that at angles of incidence that are somewhat less than the critical angle, the reflected light goes to zero. This is called the Brewster angle[16]:

$$\theta_B = \tan^{-1}\left(\frac{n_2}{n_1}\right). \quad (26)$$

The Brewster angle corresponds to the condition $\theta_1 + \theta_2 = 90°$, and Eq. 26 is obtained readily from this condition and Snell's law. The lack of reflected light at the Brewster angle occurs because the oscillators, or "antennas," that give rise to both the refracted and reflected rays cannot radiate along their electric vector axis. When the reflected ray coincides with the antennas' axis, there can be no reflected ray. Similarly, a standard antenna for a radio station has an antenna "blind cone" along the direction of the antenna. Moreover, these antennas are generally vertical to match the convenient orientation of receiver antennas for radios and the like. Many antennas on microwave towers employ the same orientation.

For a condensed phase in the gas detector flowline and with sapphire windows, the Brewster angle is ~39°. The gas detector makes a measurement of reflected light roughly spanning the critical and Brewster angles for gas (Fig. 42). This angular range corresponds to the Brewster angle for liquids, thereby removing possible light noise. Gas is indicated if a large reflection signal is obtained coinciding with the gas critical angle.

---

[16] In contrast to p-polarization, s-polarization with its electric vector transverse to the plane of incidence does not exhibit the Brewster angle phenomenon of zero reflected intensity. Instead, s-polarization exhibits a monotonic increase in reflected intensity with increasing angle of incidence. Consequently, unpolarized incident light reflecting from angled surfaces possesses much higher s-polarized intensity. Polaroid® sunglass lenses preferentially filter out s-polarization to remove the "glare," or reflected intensity, from surfaces.

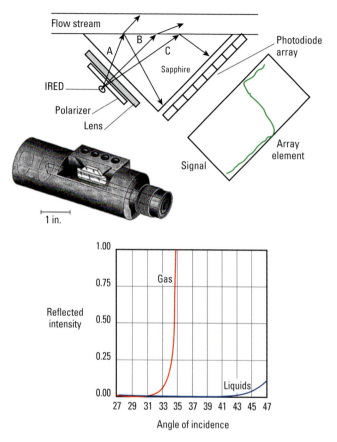

Figure 42. The gas detector employs a high-pressure cell (center). The polarizer selects the p-polarization and the cylindrical lens focuses the infrared-emitting diode (IRED) output on the flowline (top). The bottom plot shows the difference between internally reflected p-polarized light in a sapphire window for gas (at 1 atm) and water, which enables the gas versus liquid discrimination. From Mullins, Schroeder, and Rabbito (1994).

Light scattering can yield light noise, so calibration is important to discern a real gas signal. Pressurized calibration (Fig. 42) is essential (typically only a few hundred psi) to ensure proper seating of the windows and thus the optics.

Gas under borehole pressures can have indices of refraction of ~1.2, which is still much lower than that of a condensed phase but much higher than gas at 1 atm. Equation 27 gives the index of refraction for gases as a function of mass density ρ and molar (or molecular) polarizability A. At optical frequencies, the molecular polarizability is for the most part equal to the sum of the constituent atomic polarizabilities (the much slower vibrational modes and rotation modes are frozen out at the optical frequencies of ~10^{15} Hz):

$$n = \left(\frac{1+2\rho A}{1-\rho A} \right)^{1/2} \approx 1 + \frac{3}{2}\rho A . \qquad (27)$$

The shift in critical angle is

$$\Delta\theta_c \sim \sin(\theta_c^*) - \sin(\theta_c) = \frac{n_2 - 1}{1.74} , \qquad (28)$$

where θ_c^* is the critical angle at high pressure and θ_c is the critical pressure for the same gas at 1 atm (Mullins, Schroeder, and Rabbito, 1994).

For polarizable gases such as methane, there is a large change in the critical angle with pressure, whereas for helium there is almost no change. Figure 43 shows that the measured and predicted pressure dependence of the critical angle for several gases have excellent agreement (with no adjustable parameters). The measurements were made in the high-pressure gas detector cell (see Fig. 42). The molecular polarizabilities (cm^3/mole) are methane, 6.62; nitrogen, 4.44; argon, 4.19; and helium, 0.54. The extremely low polarizability of helium has to do with the fact that polarization requires an admixture of higher lying electronic states, which in helium's case are more than 20 eV above the ground state, thus there is little admixture.

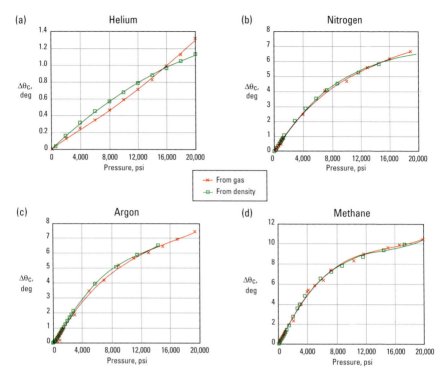

Figure 43. The index of refraction of gas depends on pressure and molar polarizability in known ways. Methane shows a large change with pressure whereas helium shows almost none. From Mullins, Schroeder, and Rabbito (1994).

This measurement is a surface, or evanescent, wave measurement, detecting gas at or near the window/flow-stream interface. The evanescent wave extends for several wavelengths of light, thus several microns beyond the interface. Flat liquid films on the window parallel to the window surface should not matter because Snell's law applies at each interface.

Fluorescence of crude oils

Downhole fluorescence

Fluorescence has been used by mud loggers ever since there were convenient sources of UV light for the purpose of examining cuttings as a direct method for detecting the presence of oil. In addition, fluorescence is visually observed, leaving a lasting impression. As such, individuals in operating companies are generally familiar with the fluorescence of crude oils. There are specific applications of fluorescence in the downhole environment. This section discusses some of the applications of downhole fluorescence. Then the interesting and dynamic science of crude oil fluorescence is treated.

Figure 44 shows several crude oils under visible light (top) and under UV light (bottom), where only fluorescence is visible (along with trace blue light from the UV lamp). The heavy oil shows little fluorescence, which is rather red. As the oils get lighter, the fluorescence intensity increases and becomes more blue. The lightest oil absorbs little of the source UV light so it shows only some fluorescence. Fluorescence *must* be initiated by photoabsorption.

Figure 44. Five crude oil samples are illuminated by visible light (top) and UV light (bottom). With the UV illumination, fluorescence is observed.

There is an additional hydrocarbon phase transition that must be detected besides gas detection. Consider a hydrocarbon in the formation at point B in Fig. 40. If the pressure is reduced on this fluid, then the dewpoint line can be intersected. The dewpoint line is the 100% gas line. If the pressure is reduced further, lines are intersected in the two-phase region that are mostly gas but with some liquid. That is, a pressure drop on this fluid results in liquid evolution. This process is opposite to expectation: pressure reduction normally yields gas, not liquid. Consequently, this process is termed retrograde condensation and is quite common in the oil field, yielding a mist of liquid droplets in a continuous gas phase. After forming, the mist often coats vessel walls and window surfaces. Furthermore, the aromatics (including resins) are concentrated in the liquid mist fraction, and it is the aromatics that are responsible for visible absorption and fluorescence emission in crude oils. The presence of retrograde dew is detected with a fluorescence measurement (Betancourt, Fujisawa, Mullins, *et al.*, 2004) using an optical configuration that has particular sensitivity to surfaces.

Fluorescence is used to detect retrograde dew transitions. The liquid phase, typically containing the color, is much more fluorescent than either the single-phase retrograde condensate or the separated gas phase. Figure 45 illustrates the detection of retrograde dew along with gas (high GOR) and water from the WBM filtrate. The fluids undergo gravity segregation in the MDT pumpout unit as a result of the relatively long residence time. Bulk gas is detected by its large methane fraction via NIR absorption, bulk liquid condensate is detected by its large fluorescence signal, and water is detected by the NIR water peak. The flowline contents are obtained by alternately draining the contents of each of the two chambers of the pumpout unit. The fluids remaining in the transfer lines are observed with each new half-cycle.

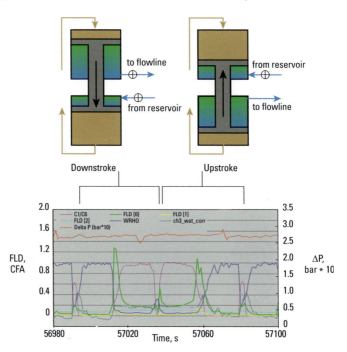

Figure 45. Fluorescence detection is used to identify retrograde dew (green trace), along with gas detection and water (blue) detection by NIR. The contents of the dead volume in the transfer lines are seen at the pump stroke turning points. From Betancourt, Fujisawa, Mullins, *et al.* (2004). © 2004 Society of Petroleum Engineers.

The identification of the fluid as a retrograde dew was easy in this case because the sampled fluid was a "gas" phase above a liquid phase in the reservoir. In addition, Fig. 45 shows that the liquid condensate phase is a small volume whereas the gas phase is a large volume. The gas is high pressure, with its density not that much lower than that of the liquid phase. This coincides with entering the two-phase region of Fig. 40 to the right of the critical point (for example, from point B). The phase curves are just below the 100% gas line, so they are in the range of 90% gas. If the two-phase region had been entered to the left of the critical point (for example, from point A), then the 100% liquid line would have been intersected and the operative phase curve could be in the range of 90% liquid. In Fig. 40, most of the volume is gas, thus the reservoir fluid is a retrograde condensate.

Another application of fluorescence is to provide oil typing when emulsions are present. Emulsions are particularly stable for heavy oil and greatly interfere with optical transmission measurements (Sjöblom, Hemmingsen, and Kallevik, 2007). As mentioned before, asphaltenes are interfacially active and are present in high concentrations in heavy oils.

Figure 46 shows the NIR spectra of several heavy oils prior to a deemulsification step (top panel) and after attempted deemulsification (bottom panel) (Andrews, Schneider, Canas, et al., 2008). Prior to deemulsification, samples D, E, and F exhibit the 1.9-um water peak associated with the emulsion. After deemulsification, samples D and E exhibit no apparent scattering and the water peak is absent. Sample F could not be deemulsified, which is often a difficult task for heavy oils. This crude oil stills exhibits the water peak at 1.9 um even after attempted deemulsification. Even for the emulsions, the water peak is small. First, it takes only a small mass fraction of emulsified water to produce strong light scattering. For example, whole milk with 4% emulsified milk fat is a much better scatterer of light (much whiter) than skim milk, which has no milk fat. In addition, the more water droplets that are intersected by the photon, the more likely it will be deflected out of the collection optics geometry.

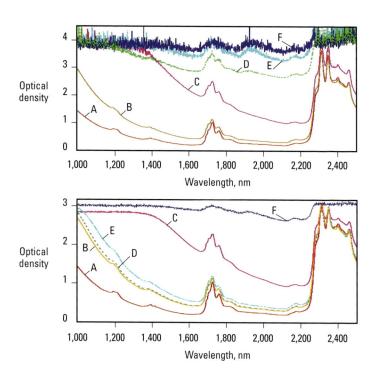

Figure 46. The NIR spectra of six heavy oil samples are shown before (top) and after (bottom) attempted deemulsification. Samples D, E, and F exhibit huge scattering in emulsified form and also exhibit the 1.9-um peak of water. Light scattering from emulsions interferes with many DFA measurements. From Andrews, Schneider, Canas, *et al.* (2008).

Figure 47 shows that fluorescence is rather independent of the state of the emulsion of the sample. The absorption length of the exciting 550-nm source light in heavy oil is very short. Only a smear of heavy oil on the window is needed to obtain a representative fluorescence spectrum. The intensity of the fluorescence can be used to obtain a qualitative indicator of oil type (compare with Fig. 44). Variation of oil type in a column (for example, induced by biodegradation) can be obtained by using DFA methods but without the pumping time required to obtain an emulsion-free sample.

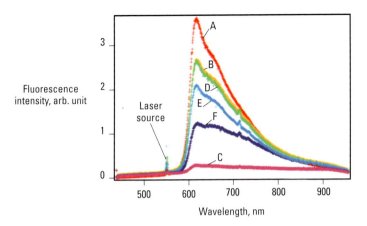

Figure 47. The fluorescence emission spectra of the emulsified crude oils of Fig. 46 are identical to those obtained for the deemulsified oils (not shown). The oil samples B and (emulsified) D have nearly the same fluorescence spectrum as expected from the NIR spectra of the deemulsified oils. From Andrews, Schneider, Canas, et al. (2008).

Emulsion formation is a common problem for heavy oils, particularly during sample acquisition for wells drilled with WBM. In addition, some degree of emulsification can occur during drilling, particularly in association with the mud spurt. The most natural form of crude oil–water emulsions is water-in-oil emulsions. These are called invert emulsions because oil-in-water emulsions are more common. Crude oils tend to form invert emulsions because the surfactants are all in the oil phase, not the water phase. A subset of asphaltenes is surfactant, thus heavy oils with their high asphaltene fraction are loaded with surfactant. Organic acids are also excellent surfactants, thus crude oils with high acid numbers also generally form stable emulsions. A subset of organic acids—tetraorganic acids—is famous for causing organic scale (Fig. 9). The stability of water-in-crude-oil emulsions is due to the following: in a mixture of crude oil and water, there is nothing to interfere with the close approach of two oil droplets. The only intervening substance is water. Consequently, oil droplets readily coalesce, thereby lowering surface energy. In contrast, the naturally occurring crude oil surfactants go to the oil/water interface. If two water droplets approach each other, the surfactant layers at each oil/water interface in the oil phase prevent the close approach of the water droplets and thus prevent coalescence. Consequently, water droplets in crude oil are stabilized. High surfactant concentrations, along with the high viscosity of heavy oils, act to stabilize heavy oil–water invert emulsions.

A very novel fluorescence application has been tested, performing a continuous fluorescence log of the oil well (Mullins, Fujisawa, Elshahawi, et al., 2005). This was attempted only in wells drilled with WBM because many of the additives in OBM fluoresce. There were difficulties associated with achieving less than 1 um of standoff from the borehole wall in the 12 wells of the field test campaign. In addition, fluorescence from limestones had to be differentiated from that of crude oil (Wang and Mullins, 1997). However, the biggest difficulty encountered by the fluorescence logging tool (FLT) (Fig. 48) was the USD 20/bbl price of oil when the tool was introduced to the field. It is unlikely that this "problem" will reoccur.

Figure 48. Emerging from a well in Patagonia, the FLT prototype was used to obtain a fluorescence log of the borehole wall (O.C. Mullins and T. Terabayashi, pers. comm., 1997).

Figure 49 shows an FLT log along with a CMR* Combinable Magnetic Resonance log and other measurements. In tight, oil-filled sands, with no invasion and no mudcake but numerous shale streaks, the FLT signal is erratic but often large. In high-permeability sands with mudcake and invasion but no shale streaks, the FLT signal is reduced but constant. The ability to cut through the mudcake and the presence of only minimal invaded mud solids are key to successful FLT logging. Other platforms for fluorescence, for example, in production logging, are rather simple to imagine.

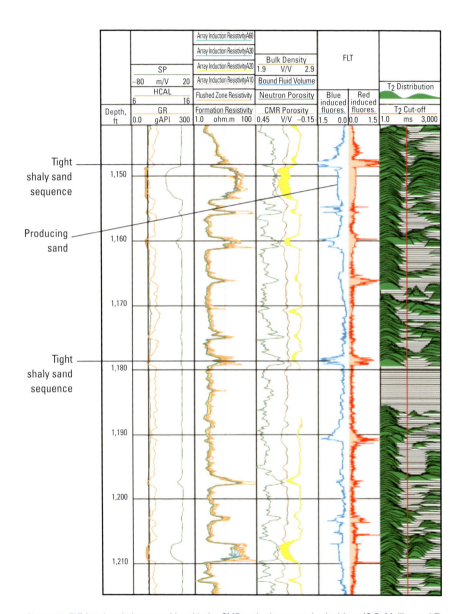

Figure 49. FLT log data is interpretable with the CMR and other petrophysical logs (O.C. Mullins and T. Terabayashi, pers. comm., 1997).

The science of crude oil fluorescence

Oil fluorescence results from the aromatic fraction of the crude oil, and different crude oils exhibit different fluorescence spectra. The aromatic fraction is not in reference to the traditional SARA (saturates, aromatics, resins, and asphaltenes) fraction designations of crude oils; in fact, all these fractions except the saturate fraction contain aromatic compounds. The fluorescence of crude oils is somewhat complicated; nevertheless, crude oil fluorescence can be treated within a single photophysical framework (Mullins, 1998). There are several excellent methods for interrogating intramolecular and intermolecular processes associated with electronic excitation. Besides optical absorption, the panoply of measurements includes fluorescence spectra, fluorescence lifetimes, and fluorescence quantum yields (intensities) as a function of concentration and crude oil type and versus excitation and emission wavelengths. First, the simple photophysical model of crude oils is described, and then the vast and wide-ranging fluorescence data of crude oil is treated. A more thorough treatment is given elsewhere (Mullins, 1998). Here, some of the more complicated formalisms are eschewed, again such as the corresponding quantum mechanics.

Upon irradiating a sample of crude oil with optical radiation of specified energy, some fluorophores are excited. As gleaned from the Urbach formalism, there are more chromophores of small size than big size (Fig. 26), and chromophore size is linked with spectral properties, with the small chromophores being bluer and the bigger ones redder (Ruiz-Morales and Mullins, 2007). Consequently, when crude oil is irradiated with a particular frequency of light, the excited fluorophore population tends to be the fraction that would emit fluorescence at nearly the same wavelength as that of the excitation light (there is a small solvent-induced red shift, the Stokes shift). Indeed, in very dilute samples, this is observed. For fluorophores A and B and quencher molecule Q, the relevant processes of excitation, emission, energy transfer (with subsequent emission), and quenching are as follows:

$A + h\nu \rightarrow A^*$	Photoexcitation of compound A to excited state A^* (absorption of light)
$A^* \rightarrow A + h\nu$	Fluorescence emission (at rate k_f)
$A^* + B \rightarrow A + B^* + \delta$	Electronic energy transfer from A to B with heat δ
$B^* \rightarrow B + h\nu'$	Fluorescence emission from B, at a longer wavelength than from A
$A^* + Q \rightarrow A + Q + \Delta$	Quenching of electronic excitation energy into heat Δ by quencher Q

The symbol * designates electronic excitation, $h\nu$ is a photon, and Δ is thermalized energy in the quenching process. The direct photoexcitation of B is not considered in this model because A is considered to dominate the photoabsorption process.

Fluorescence is not a fast process—it takes nanoseconds (Wang and Mullins, 1994). In that time, molecules can diffuse significant distances (at a molecular-length scale). The mean expectation of displacement <x> of diffusion is given in terms of time t and diffusion constant D:

$$\langle x^2 \rangle = Dt. \tag{29}$$

For typical diffusion constants of molecules in liquids of 10^{-5} cm^2/s, molecules in crude oil travel 10 Å in a nanosecond. In crude oils, which have appreciable concentrations of chromophores and fluorophores, the diffusing molecule most likely encounters a chromophore during this time and distance. For energy transfer to take place during a molecular collision, the excitation energy of the initially excited fluorophore must exceed the energy of excitation of the recipient chromophore. Energy can run only down hill. If energy transfer takes place to a fluorophore with smaller excitation energy, then the corresponding fluorescence emission resulting from the energy transfer is red shifted. In addition to energy transfer to a fluorophore of lower energy excitation, energy transfer to a nonfluorescent chromophore (a quencher) can also take place, reducing the fluorescence intensity and the fluorescence lifetime.

In crude oils, the largest chromophores and fluorophores are the asphaltenes. It has recently been established that in black oils, the asphaltenes are in nanocolloidal particles (Mullins, Betancourt, Cribbs, *et al.*, 2007; Betancourt, Ventura, Pomerantz, *et al.*, in press). The aggregation number (molecules per aggregate) is thought to be ~8 (Freed, Lisitza, Sen, *et al.*, 2007; Betancourt, Ventura, Pomerantz, *et al.*, in press; Zeng, Song, Johnson, *et al.*, in press). Consequently, in black crude oils, diffusion is not needed to cause quenching in asphaltenes. In addition, as subsequently discussed, the quantum yields of isolated asphaltene molecules are low relative to other, bluer crude oil fluorophores. Indeed, asphaltenes in crude oils decrease the fluorescence intensity. For crude oil fluorescence measurements, these bluer fluorophores, which are not asphaltenes, are of more interest.

Energy transfer effects are evident when comparing the fluorescence spectra of neat (undiluted) and very dilute crude oils. Virtually all crude oils exhibit significant (molecular) collisional energy transfer for short-wavelength excitation (Fig. 50) (Downare and Mullins, 1995). Large fluorescence red shifts are found for neat crude oils compared with dilute crude oils for short-wavelength excitation. The extent of fluorescence red shift is a function of excitation wavelength. The energy transfer is collisional, not radiative. At long-wavelength excitation, there is primarily collisional quenching at high concentrations. Consequently, at long-wavelength excitation, there is little change in the emission spectrum.

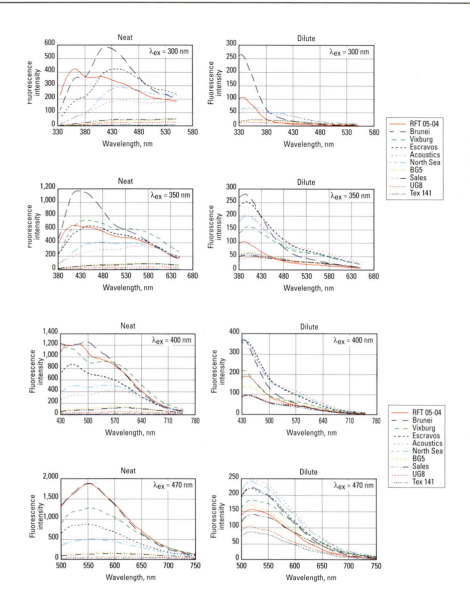

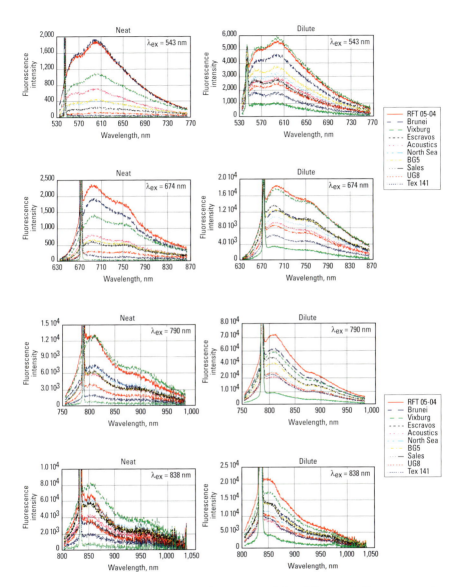

Figure 50. For short-wavelength excitation, the fluorescence emission of dilute crude oils occurs at much shorter wavelength than for neat (undiluted) crude oils. At the longest wavelength excitation, there are virtually no spectral differences for neat versus dilute crude oils. Increased noise associated with the smaller signal is generally observed for neat crude oil spectra. The oils are listed lightest to heaviest. All fluorescence intensities are listed in arbitrary units. From Downare and Mullins (1995).

Figure 51 reinforces the observations from Fig. 50, with a continuous blue shift with dilution in the fluorescence emission curve for a crude oil subjected to short-wavelength excitation. Each of these spectra is normalized to a maximum signal = 1. With increasing dilution, there is a reduced rate of collision between chromophores in solution during the lifetime of the electronically excited states responsible for fluorescence. Consequently, there is reduced collisional energy transfer with correspondingly less red shift with dilution.

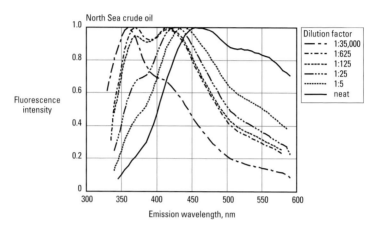

Figure 51. Fluorescence spectra of diluted crude oils show increasing blue shift of fluorescence emission with dilution. From Wang and Mullins (1994).

Collisional relaxation processes, both quenching and energy transfer, represent additional decay processes beyond intrinsic fluorescence emission and therefore yield a substantial reduction in fluorescence lifetimes (Wang and Mullins, 1994). The fluorescence quenching process in molecular collisions can be mediated through two-level quantum mechanical mixing of fluorophore and quencher electronic states (Groenzin, Mullins, and Mullins, 1999), as well as by spin-orbit coupling induced by paramagnetism (Canuel, Badre, Groenzin, *et al.*, 2003). Many crude oil spectra can be fit using a two- or three-component exponential decay curve (Fig. 52). Other fitting functions can also be used for fluorescence decay curves of crude oils (Ryder, 2005).

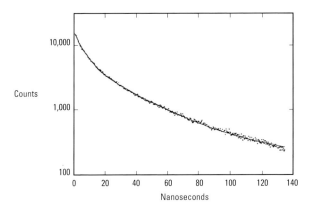

Figure 52. The fluorescence lifetime decay curves can be approximated by a two-component fit. Fluorescence lifetimes of dilute crude oils are typically on the order of 1 to 10 ns. From Wang and Mullins (1994).

The lifetime data of crude oils can be analyzed within the Stern-Volmer framework (Wang and Mullins, 1994). The decay rate of the fluorescence k_F can be expressed as a function of the intrinsic fluorescence decay rate k_{F_0} along with a collisional decay rate k_Q, which is a function of the quencher concentration [Q] and thus the crude oil concentration, with the brackets indicating molarity of molecular species Q: (For the purposes of a particular energy fluorophore, the lifetime reduction depends only on the collisional process but not whether quenching or red-shifted fluorescence takes place. Here, a quenching formalism can be used.)

$$k_F = k_{F_0} + k_Q [Q]. \qquad (30)$$

The lifetime of the excited state is reduced because of the additional decay channel, collisional quenching. Because $k_F = 1/\tau_F$ for the fluorescence lifetime τ_F, Eq. 30 can be rearranged to obtain

$$\frac{\tau_{F_0}}{\tau_F} - 1 = \frac{k_Q}{k_{F_0}} [Q]. \qquad (31)$$

Plotting the left side of Eq. 31 versus [Q] gives the Stern-Volmer plot. The same formalism can be used to obtain an expression for the reduction of fluorescence intensity. Figure 53 shows Stern-Volmer plots for lifetime and for quantum yield of North Sea crude oil.

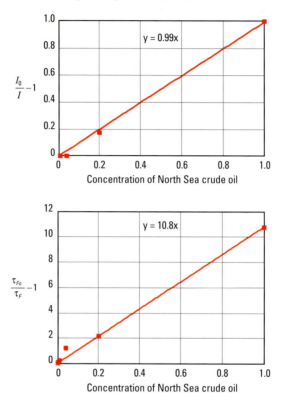

Figure 53. Stern-Volmer plots for excitation wavelength ~316 nm show that collisional interactions of chromophores in crude oils reduce fluorescence lifetimes and quantum yields. The energy transfer process is nonradiative; therefore, it affects the lifetime of the original excited molecule. The fluorescence lifetime of the originally excited fluorophores (bottom) is reduced both by quenching and energy transfer (bigger slope), whereas the fluorescence intensity (top) is reduced only by quenching (smaller slope). From Wang and Mullins (1994).

Collisional quenching of fluorescence in crude oils is indicated. The greater reduction in lifetime than intensity upon energy transfer shows that this energy transfer process is nonradiative. (In addition, the corresponding ODs are far too low for this to be radiative energy transfer [Mullins, 1998].) This holds because for a given crude oil, the lifetime plot shows a greater slope than the intensity plot; both the collisional processes of quenching and nonradiative energy transfer decrease lifetimes but only collisional quenching decreases intensity. Moreover, whether a collisional process results in quenching or energy transfer with red-shifted emission depends enormously on the excitation energy. Figure 50 shows that for short-wavelength excitation, collisional processes yield huge red shifts, whereas for long-wavelength excitation, there is almost no red shift. In the latter case, collisional quenching dominates. Why? Crude oils have been shown to adhere strictly to the Energy Gap law (Ralston, Wu, and Mullins, 1996), which is expected to apply widely but is seldom demonstrated (Turro, 1978). The Energy Gap law (Eq. 32) gives the rate of internal conversion k_{ic}, a radiationless de-excitation process that competes with fluorescence. This radiationless decay rate depends on the preexponential frequency factor A (generally about 10^{13}/s), the transition energy gap ΔE, and β, which is the vibrational energy quantum of the de-excitation mode:

$$k_{ic} = A \exp\left(-\frac{\Delta E}{\beta}\right). \tag{32}$$

The origin of the Energy Gap law rests on the fact that radiationless transitions require vibrational wave function overlap (Fig. 54). This relaxation process corresponds to the conversion of electronic excitation energy to vibrational excitation energy. The process can be thought of as inelastic electron scattering of the excited electron with the molecular core.

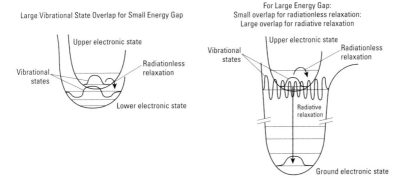

Figure 54. The Energy Gap law is applicable to crude oils. For a small energy gap, the vibrational wave function overlap is large (left), producing a facile, nonradiative relaxation. For a large energy gap, the overlap is poor (right) so fluorescence can occur. From Mullins (1998). With kind permission of Springer Science+Business Media.

For larger energy gaps, the vibrational overlap integral couples short- and long-wavelength vibrational states (Fig. 54, right), resulting in small integral values. Thus, with increasing energy gaps, radiationless transitions become increasingly forbidden. With no radiationless transition, fluorescence, an intrinsically slow process (taking nanoseconds) can occur. Nanoseconds are long compared with electronic cycle times of $\sim 10^{-16}$ s and even vibrational times of $\sim 10^{-13}$ s. In contrast, for small energy gaps, similar vibrational states are coupled, yielding large vibrational integrals and producing rapid radiationless relaxation. The quantum yield Θ for dilute crude oils is given by

$$\Theta = \frac{k_{Fo}}{k_{Fo} + k_{ic}}. \tag{33}$$

Substituting Eq. 32 into Eq. 33 gives the familiar S-curve form for the dependence of quantum yield on photon energy (Fig. 55). The band-gap ΔE in Eq. 32 defines the fluorophores in crude oil that can be excited by that photon energy.

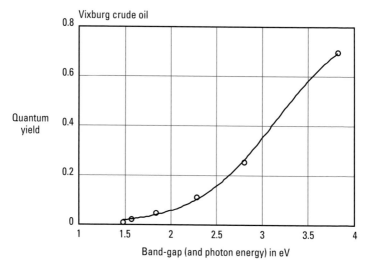

Figure 55. The quantum yield versus band-gap of crude oils obeys a curve derived from the Energy Gap law, which dominates the fluorescence properties of crude oils. From Ralston, Wu, and Mullins (1996).

The range of electronic band-gap transition energies of crude oil extends into the NIR in Fig. 55. This band-gap is also called the transition from the highest occupied electronic state to the lowest unoccupied electronic state (HO-LU gap). The close fit of the quantum yield data over a broad energy range to the Energy Gap law proves the law's validity. In fact, crude oils are one of the few systems that are shown to fit this well-known law. Most systems do not have a continuously variable band-gap as does the ensemble of crude oil fluorophores. Moreover, because crude oils contain a large number of fluorophores, they are not subject to the idiosyncrasies of individual fluorophores. The excitation energy dependence of the quantum yield in Eqs. 32 and 33 has been shown to apply to all crude oils studied (Ralston, Wu, and Mullins, 1996) albeit dilute solutions of heavier crude oils exhibit significantly lower intrinsic quantum yields than 1. For any given excitation wavelength, the fluorescence properties are dominated by fluorophores with their HO-LU gap at this energy; larger HO-LU gap molecules are, of course, not excited, and smaller HO-LU gap molecules are much less numerous, as shown by Urbach tail behavior.

The decay parameter obtained for crude oils is 3,500 cm^{-1}, which probably corresponds to de-excitation by coupling with aromatic CH stretch modes ($E \sim 3{,}050$ cm^{-1}). Although vibronic coupling between excited π states and CH stretch modes can be weak, the small number of vibrational quanta involved for this high-frequency mode increases the probability of vibronic coupling. In addition, the frequency factor of 10^{13} Hz for this vibration helps put this mode into consideration for de-exciting a nanosecond process.

The Energy Gap law coupled with the Rydberg equation explains why fluorescence generally occurs only from the first electronically excited state (within the singlet spin manifold). From the Rydberg equation, Eq. 34 is obtained to give the energy difference between the electronic levels of the principal quantum numbers n and n':

$$\Delta E_{nn'} = R\left(\frac{1}{n^2} - \frac{1}{n'^2}\right), \tag{34}$$

where R is the Rydberg constant of 13.6 eV. The largest energy gap $\Delta E_{nn'}$ between adjacent levels is between the ground ($n = 1$) and first excited state ($n = 2$). Other electronic states are much closer in energy and can undergo efficient radiationless relaxation. Molecular electronic states deviate substantially from the Rydberg equation; nevertheless, they share the overall traits of this equation. The big energy jump is the first one. De-exciting molecules often get stuck in the first excited state if the energy gap is sufficient and can then fluoresce.

Crude oils fluoresce when they are hot (Zhu and Mullins, 1992), thus fluorescence techniques can be used in situ in oil wells, where elevated temperatures are the norm. Crude oils exhibit thermally activated intersystem crossing (given by the Boltzmann factor, Eq. 11) as many organics do, but the reduction in fluorescence intensity is large only for oils that have very high quantum yields to begin with, which are the light crude oils. Apparently the labile fluorophores have already de-excited in heavy crude oils, so thermally activated intersystem crossing has a negligible effect in heavy oils (Zhu and Mullins, 1992).

The huge wavelength dependence of the fluorescence red shift on excitation wavelength in crude oils is now understood (Fig. 50). For crude oils with short-wavelength excitation, chromophore collisional processes are dominated by energy transfer with fluorescence emission. For crude oils with long-wavelength excitation, chromophore collisional processes are dominated by quenching. These observations are dictated by the Energy Gap law. At high excitation energy, excited fluorophores are likely to collide with chromophores with a large band-gap and thus high quantum yield. The Urbach population stipulates that there are many more small chromophores (with a large band-gap) than large chromophores (with a small band-gap). At low energies of excitation, the small band-gap molecule can pass energy only to even larger chromophores (with a smaller band-gap). The Energy Gap law stipulates that these molecules have very small quantum yields. By comparing the spectra of neat and dilute solutions of a given crude oil, the fraction of energy transfer (with subsequent fluorescence emission) can readily be determined (Ralston, Wu, and Mullins, 1996). Figure 56 shows that the magnitude of the energy transfer matches the quantum yield for crude oils. The Energy Gap law is now understood to dominate many of the fluorescence properties of crude oils. Because the fluorescence intensity and spectra of neat crude oils are a strong function of oil type, downhole fluorescence is useful for this characterization.

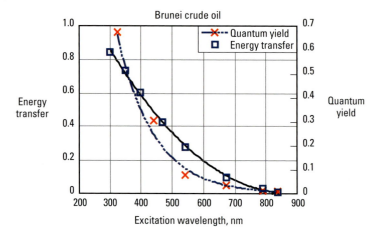

Figure 56. In a collisional process of crude oil chromophores, the quantum yield (dictated by the Energy Gap law) determines whether fluorescence from energy transfer or quenching results. From Ralston, Wu, and Mullins (1996).

Transition zone characterization and downhole pH measurement

DFA has many uses for development where the production of oil is ongoing. For example, in many of the large Middle East carbonate fields that have been producing oil for a long time, production is getting more complicated. In particular, much of the remaining producible oil is in the transition zone, where both oil and original water (connate water) are mobile. (Connate water is to be distinguished from injected water, which is used to sweep oil out of the reservoir.) In simple cases, there is an oil zone at the top of the column that produces pure oil, a middle transition zone with mobility of both connate water and oil, and a water zone below. Characterization of transition zones is problematic for many reasons, including WBM invasion, pressure depletion, capillary pressure effects, variability in rock wettability, and, of course, the arrival of the water sweep front (Carnegie, 2006). Moreover, what really matters is not only the fluids contained in the formation but rather what fluids flow out of the formation. Of course, DFA can differentiate water from oil to determine volumetric flows. However, formation fluids must be differentiated from drilling fluid filtrate. For land and shallow-water environments, WBM is often the drilling fluid of choice because of its lower cost and reasonable effectiveness. Typically, both WBM filtrate and connate water contain significant dissolved salt, so conductivity is not very effective at this differentiation. Thus, DFA must be used to differentiate WBM filtrate from connate water—but what analyte?

Litmus paper has been used for many centuries for the colorimetric determination of pH. Frequently, WBM is very basic, with a pH in the 8 to 10 range, and much more basic than the connate water. The concept employed to measure pH downhole is to inject a pH dye into the water flow stream coming from the formation and to measure the resulting color. A procedure has been developed to transport a pH dye solution downhole and inject small quantities of it into the flowline of the sampling tool (Raghuraman, O'Keefe, Eriksen, *et al.*, 2005; Raghuraman, Xian, Carnegie, *et al.*, 2005). Measuring pH with this procedure enables the differentiation of WBM filtrate from connate water. Moreover, pH measurement of formation water samples in the laboratory is problematic. With temperature reduction, pH-active salts such as $CaCO_3$ can precipitate from solution, changing the pH. In addition, the water samples are often flashed, releasing pressure and venting evolved gas. In this process, CO_2 can escape, again changing the pH of the water sample. Thus, for a variety of reasons, it is desirable to measure pH downhole. Potentiometric methods for measuring pH downhole have been attempted, but they suffer from various effects, such as surface charging of the requisite pressuring housings, streaming potentials, drift, and corrosion of electrodes, let alone fouling from crude oil. The robust optical measurements provide the best method for pH determination.

pH is defined as

$$pH = -\log[H^+]. \qquad (35)$$

The "p" means "negative log of," so pH is the negative log of [H⁺]. This log is base 10, not the natural log base e. The brackets indicate molarity. Thus, pH means the negative log of the hydrogen ion concentration.

For water at 298 K, the dissociation constant is

$$H_2O \leftrightarrows H^+ + OH^-; K_w = [H^+][OH^-] = 10^{-14}, \tag{36}$$

where again the units of [H⁺] and [OH⁻] are molarity. Thus, for pure water, [H⁺] = [OH⁻] = 10^{-7} molar; thus, the pH of pure water is 7.

The effects of ionic strength, temperature, and pressure on the relevant equilibrium constants for pH have been determined (Raghuraman, Gustavson, Mullins, et al., 2006). Measuring pH at elevated temperatures requires knowing the temperature dependence of the equilibrium constants. In particular, for a given pH-sensitive dye, acid form A and base form B⁻ have different and resolvable optical absorption spectra:

$$A \leftrightarrows B^- + H^+. \tag{37}$$

Figure 57 shows the colorimetric analysis of pH with the easily resolvable acid and base spectral peaks. The pH is given by

$$pH = pK_a + \log\left(\frac{\gamma_B}{\gamma_A}\right) + \log\frac{[B^-]}{[A]}. \tag{38}$$

where K_a is the acid dissociation equilibrium constant and the γ_i parameters are activity coefficients that account for the nonideal behavior of chemical mixtures. The equation is more commonly written

$$pH = pK'_a + \log\frac{[B^-]}{[A]}, \tag{39}$$

where the activity coefficients are incorporated into the equilibrium constant. Thus K_a' is a function of ionic strength in addition to temperature and pressure. These functional dependencies have been determined for the pH dyes of interest, which must be able to survive elevated temperatures for one day to one week depending on the application (Raghuraman, Gustavson, Mullins, *et al.*, 2006).

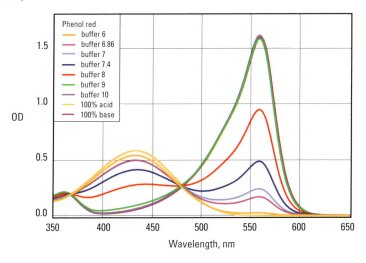

Figure 57. Similar to litmus paper, a pH dye has different colors for its acid and base form. Optical spectra can be used to determine the relative fractions of the acid and base forms, thereby giving pH. From Xian, Raghuraman, Carnegie, *et al.* (2007).

pH is not a linear function of WBM filtrate contamination in connate water. Moreover, the H^+ concentration is not even a conserved quantity. So the question arises whether the pH monitoring of water sample cleanup actually works. To answer this question, a tracer was added to a mud system to monitor WBM filtrate contamination. The tracer needed to be detectable within the measurement suite provided by existing DFA tools. Because the pH optics technology is compatible with blue dye tracers, a "Schlumberger blue" dye colorant was added to the WBM system (Hodder, Samir, Holm, *et al.*, 2004; Mullins, Hodder, Ayan, *et al.*, 2004). Comparison of pH measurement with the blue dye method of monitoring the water cleanup finds close agreement for the contamination measurements (Fig. 58) (Raghuraman, Xian, Carnegie, *et al.*, 2005). Because this agreement has been established in the field, some operating companies now use the simpler pH measurement for WBM contamination monitoring.

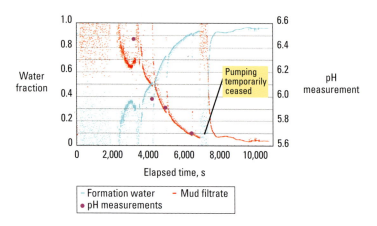

Figure 58. DFA distinguishes WBM filtrate from formation water by measurement of pH and by determination of the concentration of blue dye added to the mud. From Raghuraman, Xian, Carnegie, et al. (2005). © 2005 Society of Petroleum Engineers.

Chain of custody

Forensic science requires the establishment of a chain of custody for crime scene evidence to be used in a court of law. The chemical waste disposal industry is subject to regulations mandating chain of custody of the chemical waste. Surprisingly, in the oil industry, major economic decisions are routinely based on results from laboratory analyses of hydrocarbon fluids without any chain of custody measurements. The question arises: What is the relationship of the laboratory report to the reservoir fluids? Openhole sample acquisition and DFA are routine for obtaining information at an early time regarding reservoir fluids. A variety of problems can arise during openhole sample acquisition, but problems are not limited to open hole. To avoid phase changes and sample contamination, DFA methods are helpful.

Obtained downhole, the sample is placed in a high-pressure sample bottle and brought to the surface at the wellsite. The sample cools at surface temperatures, and if pressurized bottles are not used, the pressure drops significantly. Asphaltene flocculation can occur, particularly with a pressure drop, and wax precipitation can occur, particularly with a temperature drop. Gas evolution or retrograde dew evolution can occur with a drop in pressure and temperature. If necessary, downhole samples are transferred on the rig to government-certified transportable high-pressure bottles. (Some downhole bottles are transportable, obviating the need for sample transfer.)

For rig-site or any other sample transfer, the sample is (ideally) reconstituted to single-phase conditions. The sample is then shipped to a laboratory, and typically not until many months later is the sample reconditioned to single phase and transferred to various laboratory analysis cells for fluid properties evaluation.

A myriad of problems exists in this process. Samples might not be reconstituted to single phase at sample transfer, generally unbeknownst to the engineer. Transfer of a two-phase sample destroys the integrity of the sample; typically only part of the sample is transferred because often there are water, mud, and solids in the bottom of the bottle. Any sample bottle can leak at any point in this chain. If sample leakage occurs from a two-phase sample, then the integrity of the sample is destroyed. For instance, if gas leaks from a high-pressure bottle, then the laboratory measurement of GOR could be quite precise but not that of the reservoir fluid. Or course, errors in analysis can easily occur in a laboratory. Finally, mislabeling of bottles or samples is not uncommon at any point in the chain. Production engineers often state that laboratory reports *never* match the properties of the produced fluids.

Chain of custody of the hydrocarbon fluids is evidently needed, and DFA provides the opportunity to do this (Betancourt, Bracey, Gustavson, *et al.,* 2006). DFA is performed on the reservoir fluid after minimal sample handling and virtually no storage time. The industry's first chain of custody studies compare the DFA optical spectrum of the crude oil obtained downhole with the optical spectrum of the same sample in the laboratory when the laboratory studies are conducted. If the laboratory and downhole optical spectra match, then the laboratory report is validated. Conversely, if the two spectra do not match, then the source of error must be uncovered. No interpretation is needed in this comparison because the compared physics response is of a measurement on the fluid downhole and on the fluid in the laboratory. Discrepancies in DFA versus the laboratory determination of derived parameters such as GOR can be traced to invalid transfers, invalid algorithms, DFA tool problems, etc. For example, Fig. 59 compares DFA and laboratory spectral measurements, with excellent agreement obtained. The slight difference in the methane channel absorption between the downhole and the laboratory spectra was traced to an improper sample transfer protocol, which has since been fixed. Because production problems are often encountered years after openhole sample acquisition, it is imperative to acquire chain of custody validation at the time of sample analysis, not several years later when production starts. By then it is too late. Chain of custody is now commercial and should become standard practice.

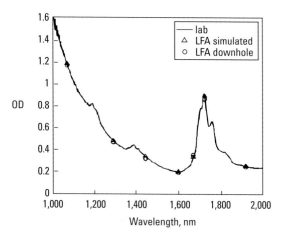

Figure 59. Comparison of the downhole and laboratory spectra of live reservoir fluid confirms the downhole data and validates the laboratory sample analysis. From Betancourt, Bracey, Gustavson, et al. (2006).

DFA fluid profiling: A quasi-continuous downhole fluid log

Cost efficiency: Toward a continuous fluid log

DFA is an essential tool for understanding fluid and reservoir complexities. Moreover, DFA is central to the process of efficient analysis of hydrocarbons (Hashem, Elshahawi, and Ugueto, 2004; Elshahawi, Venkataramanan, McKinney, et al., 2006; Mullins, Elshahawi, Hashem, et al., 2005). Essentially, without DFA there is no informed real-time decision making; samples are simply acquired from various points and laboratory analyses performed. Without DFA, uncovering reservoir fluid complexities is severely impaired. Sample acquisition and analysis are expensive, and without real-time DFA proving the existence of fluid complexities, it is difficult to justify ample sample acquisition with laboratory analysis. DFA enables matching the complexity of the sample acquisition job to the complexity of the fluids in the formation. Simple fluid columns mandate minimal DFA and sample acquisition programs; complex fluid columns mandate extensive DFA and sample acquisition programs.

The trend is now established of making multiple DFA measurements throughout an oil column, essentially constructing a "continuous" fluid log (Hashem, Elshahawi, and Ugueto, 2004; Elshahawi, Venkataramanan, McKinney, et al., 2006; Mullins, Elshahawi, Hashem, et al., 2005). Each DFA measurement is made at a discrete depth station, but with sufficient coverage a quasi-continuous fluid log is developed. Proper statistical analyses must be performed on the corresponding DFA results in order to know whether the differences are significant (Venkataramanan, Fujisawa, Mullins, et al., 2006). In this manner, the fluid properties provide information not only about their own complexities but also about the complexities of the reservoir. Many of the DFA stations obviate the need for sample acquisition. At the time of writing, typical jobs include 15 DFA stations per well with only few sample acquisition stations. Indeed, the vision of obtaining many DFA stations to ferret out fluid and reservoir complexities is being achieved. In three very different producing environments spanning high cost to low cost, there have been jobs with more than 30 DFA stations per well. The fluid complexities justified this similar DFA coverage within the different cost constraints. The vision is correct, and efficient formation evaluation mandates implementation of this vision globally on all current and future DFA platforms.

The protocol is rather straightforward for optimal DFA application; a high density (large number per unit depth) of DFA station measurements is obtained around potential permeability barriers where fluid properties may change discontinuously. A high density of DFA station measurements should also be obtained near potential fluid phase changes and contacts. Finally, a high density of DFA station measurements is recommended for fluid columns that exhibit rapid variation with depth. Everywhere else, a reduced density of DFA station measurements can be used. The actual number of measurements is tempered by the risk tolerance for errors in evaluation of the formation and reservoir fluid. For low risk tolerance, higher numbers of DFA stations are employed. The expectation is that because the easy reservoirs have largely been drained and more complex, expensive reservoirs are becoming prime targets, risk should be reduced with larger numbers of DFA stations. Moreover, as the goals of the operating companies evolve toward more efficient reservoir exploitation, as opposed to low-cost but inefficient operations, the application of DFA will naturally expand. In addition, acquiring a large number of DFA stations guides laboratory analysis of the samples. Only samples that are distinct merit acquisition for subsequent laboratory analyses, and typically only a few stations are so processed.

Petroleomics, the future of hydrocarbon analysis

In the past, the analysis of hydrocarbon gases and light-end analysis of crude oils have been proper within a strict chemical framework. For example, the complete chemical composition of a hydrocarbon gas would be accomplished in part to identify any significant production issues. Of course, the energy content of a gas establishes its economic value. The possible production of any condensate affects facilities design. The concentration of deleterious components such as H_2S and CO_2 must be known. There has not been debate as to the value of such analysis; it is presumed to be obvious. In stark contrast, heavy-end analysis of crude oils has never been proper in any chemical sense. Furthermore, in part owing to atavistic comfort with ineffective but routine methods of the past, there has even been debate as to the value of such proper analysis.

The chemical analysis of asphaltenes, the most enigmatic component of crude oils and a major constituent of heavy ends, has been pernicious and problematic (Mullins, Sheu, Hammami, *et al.*, 2007; Groenzin and Mullins, 1999). In particular, debate has raged about the fundamental properties of asphaltenes, including molecular weight, molecular architecture, nanoaggregate formation, clustering, chemical identity of interfacially active components, and reversibility of flocculation. However, many of these debates have been resolved by the innovative exploitation of advanced chemical methods. The time has come to expand our horizons.

The objective of predictive petroleum science is within sight; this is the vision of the new field of petroleomics. To perform predictive science, it is a requirement to know the structure of the system. Without knowing the structure, predictive science is precluded and the investigation is limited to phenomenology. Francis Crick, Nobel laureate for deciphering the structure of DNA, advises, "to understand function, study structure" (Crick, 1990). For crude oil, "structure" means understanding all chemical constituents, especially the asphaltenes, and understanding the hierarchical aggregate structures in crude oil. The centroid and distribution of asphaltene molecular weight have been sorted out (Mullins, Sheu, Hammami, *et al.*, 2007). This development is essential to perform heavy-end chemical analysis. On another front, the existence of asphaltene nanoaggregates has been established in the laboratory, and they are now known to exist in reservoir crude oil. Indeed, this advance has enabled the identification of compartmentalization and reservoir connectivity in case studies (Mullins, Betancourt, Cribbs, *et al.*, 2007). Recent advances in mass spectroscopy have enabled obtaining a complete listing of the elemental composition of each polar component in a heavy oil—this listing is designated the petroleome, in relation to the genome of genomics (Marshall and Rodgers, 2004; Mullins, Sheu, Hammami, *et al.*, 2007). The current mass accuracy and mass resolution are sufficient to obtain a unique listing of the elemental constituents making up the mass spectral peaks of crude oils and asphaltenes. In addition, two-dimensional gas chromatography (GC×GC) yields chemical specificity of the components of crude oils (Reddy, Nelson, Sylva, *et al.*, 2007). GC×GC analyzes the nonpolars whereas ultrahigh-resolution mass spectroscopy is best applied to polars and aromatics; the two methods are very complementary. Thus, petroleomics is enabled as a new field that embodies performing proper predictive petroleum science (Marshall and Rodgers, 2004; Mullins, Sheu, Hammami, *et al.*, 2007). This is not only a shared industry vision, with several operating companies now having their own petroleomics programs, but also a partnership of industry, university, and government scientists. Indeed, the future is very bright in this potent field.

The interface of petroleomics and DFA is currently being developed. Both DFA and petroleomics represent enormous technical advances. Linking these together into a single vision promises to improve significantly the efficiency of reservoir exploitation. Figure 60 shows a typical ultrahigh-resolution mass spectrum of a crude oil. The specific mass spectral technique is electrospray ionization Fourier transform ion cyclotron resonance mass spectroscopy (ESI FT-ICR MS).

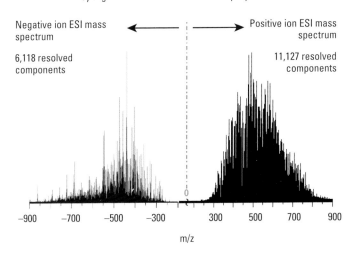

Figure 60. Positive and negative electrospray ionization has excellent resolving power. The 9.4-tesla Fourier transform ion cyclotron resonance mass spectra of a crude oil enables resolution of the basic (right) and acidic (left) components. From Rodgers and Marshall (2007). With kind permission of Springer Science+Business Media.

Each nuclide has a unique weight. Consequently, with sufficient resolving power, each peak in the mass spectrum of Fig. 60 can be assigned a unique chemical composition, $C_cH_hN_nO_oS_s$. The resolving power depends on the magnetic field strength, and significant improvements continue to be made. Chemical classes of compounds in crude oil can then be represented in plots (Fig. 61). For example, a chemical class can be formed of all compounds containing a single basic nitrogen atom. In the acquisition of a mass spectrum, the ratio of charge to mass is determined. Here, basic nitrogen is protonated to give the charge.

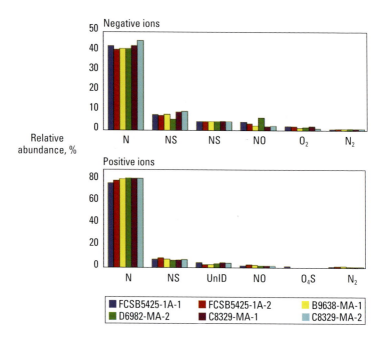

Figure 61. The mass spectral results are plotted by specific heteroatom[17] content, compounds that contain a single nitrogen atom, compounds that contain a single oxygen atom, etc. These chemical classes are plotted for four different MDT crude oil samples F, B, D, and C and replicate runs for samples F and C, corresponding to different MDT sample bottles collected at the same station. From Mullins, Rodgers, Weinheber, et al. (2006). © 2006 American Chemical Society.

Any given chemical class can be represented in terms of the carbon number, which is the number of carbon atoms in the compound, and in terms of its double bond equivalent (DBE), which is largely a measure of the aromatic carbon content. Figure 62 shows corresponding DBE versus carbon number plots for the basic nitrogen of the six crude oil samples in Fig. 61.

---

[17]*A heteroatom (from the ancient Greek* heteros *for different and* atomos*) is any atom that is not carbon or hydrogen. It is typically, but not exclusively, nitrogen, oxygen, sulfur, vanadium, or nickel in crude oil.*

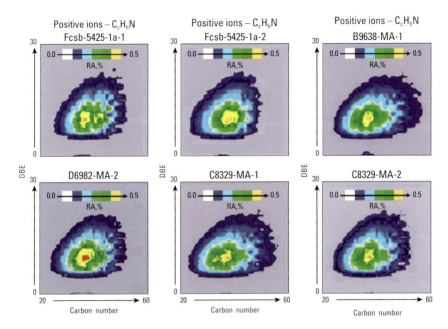

Figure 62. Percent relative abundance is plotted for the DBE (essentially aromaticity) versus carbon number for the compounds in the six MDT crude oil samples in Fig. 61. These compounds contain a single pyridinic nitrogen atom as the only heteroatom. From Mullins, Rodgers, Weinheber, et al. (2006). © 2006 American Chemical Society.

A limitation of mass spectroscopy in general is its difficulty in obtaining the alkane fraction. Mass spectroscopy measures the charge to mass ratio, so ionization is fundamental to the process. The described studies were performed with ESI, which won a Nobel Prize in 2002 for John Fenn for ionization without fragmentation of very large molecules. This capability is essential in petroleum analysis. However, alkanes are notoriously difficult to ionize by ESI because they contain no heteroatoms and thus no sites of charge separation. Other ionization methods, such as electron impact, fragment the compounds, precluding the analysis of complex mixtures.

As previously mentioned, a complementary analytical method to ultrahigh-reslution mass spectroscopy is GC×GC (Reddy, Nelson, Sylva, et al., 2007). Standard GC performs some separation, particularly of light ends, but the separation of large compounds is problematic because of overlapping signals. In GC×GC, a second GC column is employed in series; for example, separation can be obtained in the first column based on molecular size and in the second column based on polarizability. Figure 63 shows a typical GC×GC for an oil sample. The exquisite molecular resolution is evident. Different detectors can be used, such as a flame ionization detector (FID) or even a mass spectrometer (MS). Biodegradation and water washing are readily analyzed by GC×GC. Because asphaltenes do not move through a GC column, they are not analyzed by GC×GC. But asphaltenes are readily analyzed by mass spectroscopy. Linking GC×GC with DFA to enable using these advanced analytical methods to address the reservoir is in progress (Mullins, Ventura, Nelson, et al., 2008; Betancourt, Ventura, Pomerantz, et al., in press).

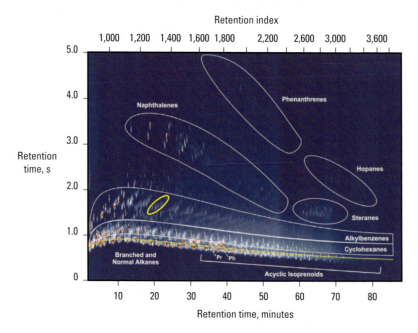

Figure 63. This GC×GC-FID chromatogram is of crude oil sample F-1 acquired with the MDT tool. The outlined areas represent peaks of specific compound classes. The yellow line marks the elution of *n*-alkanes and branched alkanes. Pr and Ph are respectively the compounds pristane and phytane. The yellow circle identifies the area analyzed by GC×GC-time-of-flight (TOF) MS. The OBM filtrate composed of monoalkenes is evident between the alkanes and cycloalkanes. After Reddy, Nelson, Sylva, *et al.* (2007).

Spider diagrams can be prepared to determine the similarity of crude oil samples. Figure 64 shows a comparison of four MDT crude oil samples from two different reservoirs. The two PER crude oil samples are seen to be identical and the two F samples are as well, but they are distinct from each other.

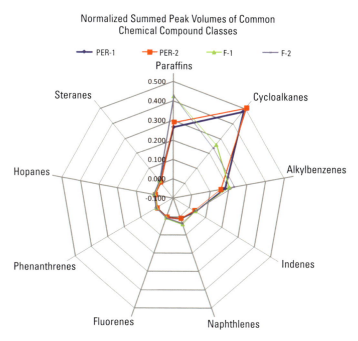

Figure 64. A spider diagram was constructed of the normalized, summed integrated peak volumes contributing to chemical compound classes after the contributions of drilling fluid contamination were removed. The axis scale is adjusted to −1.0 to better emphasize the distributions of low-yield compound classes. From Betancourt, Ventura, Pomerantz, et al. (in press). © 2008 American Chemical Society.

These new laboratory capabilities are finally enabling a detailed chemical characterization of crude oil. The vision of petroleomics, relating properties to structure (and thus chemical composition), is now being realized. Of course, flow assurance can be addressed. However, the biggest problem by far in the oil industry today is understanding the reservoir. Clearly, petroleomics should address the reservoir; petroleomics studies of the reservoir necessarily start with DFA. That is, DFA places the crude oil chemistry in the context of the reservoir. A partnership across corporate, university, and national laboratory boundaries is required to link DFA with leading analytical chemistry methods to address the reservoir.

Conclusions

Downhole fluid analysis has rapidly become a key method for understanding the reservoir. Two simplifying assumptions regarding reservoirs are increasingly viewed as invalid and thus costly: that reservoirs have a single giant tank filled with one homogeneous hydrocarbon. DFA has been central in elucidating the reality of reservoir complexities, of both the reservoir fluids and reservoir architecture. In particular, DFA naturally identifies spatial fluid compositional variation. In addition, reservoir architecture affects the dynamic processes involving reservoir fluids that create fluid variations; therefore, DFA is quite sensitive to this structure. Moreover, DFA enables matching in real time the cost of the DFA job to the complexities of the reservoir within the corresponding cost-benefit considerations.

Essentially DFA corresponds to chemical analysis. Separation science is the routine chemical analysis procedure for the upstream oil business; however, DFA is performed within a setting of a process stream. Spectral analysis is routine in process streams such as those in refineries. Because DFA methods are not common in the upstream oil business and, moreover, because the DFA setting offers some unique challenges, it is valuable to document DFA-related physics in a single place. That is the purpose of this chapter. Future DFA methods will broaden considerably beyond those presented herein. For the foreseeable future, these new methods will supplement, not supplant, optical spectroscopy. Most likely, the materials in this book and the time spent deciphering its contents will remain valuable.

References

Alpak, F.O., Elshahawi, H., Hashem, M., and Mullins, O.C.: "Compositional Modeling of Oil-Based Mud-Filtrate Cleanup During Wireline Formation Tester Sampling," paper SPE 100393 presented at the SPE Annual Technical Conference and Exhibition, San Antonio, Texas, USA (September 24–27, 2006).

Andrews, A.B., Guerra, R.E., Mullins, O.C., and Sen, P.N.: "Diffusivity of Asphaltene Molecules by Fluorescence Correlation Spectroscopy," *Journal of Physical Chemistry A* (2006) 110, No. 26, 8093–8097.

Andrews, A.B., Schneider, M.H., Canas, J., Freitas, E., Song, Y., and Mullins, O.C.: "Fluorescence Methods for Downhole Fluid Analysis of Heavy Oil Emulsions," *Journal of Dispersion Science and Technology* (February 2008) 29, No. 2, 171–183.

Atkins, P.W., and Friedman, R.S.: *Molecular Quantum Mechanics* (4th ed.), Oxford, England, Oxford University Press (2005).

Bergmann, U., Groenzin, H., Mullins, O.C., Glatzel, P., Fetzer, J., and Cramer, S.P.: "Carbon K-Edge X-Ray Raman Spectroscopy Supports Simple yet Powerful Description of Aromatic Hydrocarbons and Asphaltenes," *Chemical Physics Letters* (February 2003) 369, Nos. 1–2, 184–191.

Betancourt, S.S., Bracey, J., Gustavson, G., Mathews, S.G., and Mullins, O.C.: "Chain of Custody for Samples of Live Crude Oil Using Visible-Near-Infrared Spectroscopy," *Applied Spectroscopy* (December 2006) 60, No. 12, 1482–1487.

Betancourt, S.S., Fujisawa, G., Mullins, O.C., Eriksen, K.O., Dong, C., Pop, J., and Carnegie, A.: "Exploration Applications of Downhole Measurement of Crude Oil Composition and Fluorescence," paper SPE 87011 presented at the SPE Asia Pacific Conference on Integrated Modelling for Asset Management, Kuala Lumpur, Malaysia (March 29–30, 2004).

Betancourt, S.S., Ventura, G.T., Pomerantz, A.E., Viloria, O., Dubost, F.X., Zuo, J., Monson, G., Bustamante, D., Purcell, J.M., Nelson, R.K., Rodgers, R.P., Reddy, C.M., Marshall, A.G., and Mullins, O.C.: "Nanoaggregates of Asphaltenes in a Reservoir Crude Oil," *Energy & Fuels* (in press).

Brown, C.W., and Lo, S.-C.: "Spectroscopic Measurement of NaCl and Seawater Salinity in the Near-IR Region of 680–1230 nm," *Applied Spectroscopy* (January 1993) 47, No. 2, 239–241.

Canuel, C., Badre, S., Groenzin, H., Berheide, M., and Mullins, O.C.: "Diffusional Fluorescence Quenching of Aromatic Hydrocarbons," *Applied Spectroscopy* (May 2003) 57, No. 5, 538–544.

Carnegie, A.J.G.: "Understanding the Pressure Gradients Improves Production from Oil/Water Transition Carbonate Zones," paper SPE 99240 presented at the SPE/DOE Symposium on Improved Oil Recovery, Tulsa, Oklahoma, USA (April 22–26, 2006).

Cooper, J.B., Wise, K.L., Welch, W.T., Sumner, M.B., Wilt, B.K., and Bledsoe, R.R.: "Comparison of Near-IR, Raman, and Mid-IR Spectroscopies for the Determination of BTEX in Petroleum Fuels," *Applied Spectroscopy* (November 1997) 51, No. 11, 1613–1620.

Crick, F.: *What Mad Pursuit: A Personal View of Scientific Discovery*, New York, New York, USA, Basic Books (1990).

Del Campo, C., Dong, C., Vasques, R., Hegeman, P., and Yamate, T.: "Advances in Fluid Sampling with Formation Testers for Offshore Exploration," paper OTC 18201 presented at the Offshore Technology Conference, Houston, Texas, USA (May 1–4, 2006).

Dong, C., Hegeman, P.S., Mullins, O.C., Fujisawa, G., Betancourt, S.S., Pop, J., Kurkjian, A.L., Terabayashi, T., and Elshahawi, H.M.: "Determining Fluid Properties from Fluid Analyzer," US Patent No. 6,956,204 (October 18, 2005).

Dong, C., Mullins, O.C., Hegeman, P.S., Teague, R., Kurkjian, A., and Elshahawi, H.: "In-Situ Contamination Monitoring and GOR Measurement of Formation Fluid Samples," paper SPE 77899 presented at the SPE Asia Pacific Oil and Gas Conference and Exhibition, Melbourne, Australia (October 8–10, 2002).

Dong, C., O'Keefe, M., Elshahawi, H., Hashem, M., Williams, S., Stensland, D., Hegeman, P., Vasques, R., Terabayashi, T., Mullins, O., and Donzier, E.: "New Downhole Fluid Analyzer Tool for Improved Reservoir Characterization," paper SPE 108566 presented at Offshore Europe, Aberdeen, Scotland, UK (September 4–7, 2007).

Downare, T.D., and Mullins, O.C.: "Visible and Near-Infrared Fluorescence of Crude Oils," *Applied Spectroscopy* (June 1995) 49, No. 6, 754–764.

Ellis, D.V., and Singer, J.M.: *Well Logging for Earth Scientists* (2nd ed.), Dordrecht, The Netherlands, Springer (2007).

Elshahawi, H., Venkataramanan, L., McKinney, D., Flannery, M., Mullins, O.C., and Hashem, M.: "Combining Continuous Fluid Typing, Wireline Formation Tester, and Geochemical Measurements for an Improved Understanding of Reservoir Architecture," paper SPE 100740 presented at the SPE Annual Technical Conference and Exhibition, San Antonio, Texas, USA (September 24–27, 2006).

Freed, D.E., Lisitza, N.V., Sen, P.N., and Song, Y.-Q.: "Molecular Composition and Dynamics of Oils from Diffusion Measurements," *Asphaltenes, Heavy Oils, and Petroleomics*, O.C. Mullins, E.Y. Sheu, A. Hammami, and A.G. Marshall (eds.), New York, New York, USA, Springer (2007), 279–300.

Fujisawa, G., Betancourt, S.S., Mullins, O.C., Torgersen, T., O'Keefe, M., Terabayashi, T., Dong, C., and Eriksen, K.O.: "Large Hydrocarbon Compositional Gradient Revealed by In-Situ Optical Spectroscopy," paper SPE 89704 presented at the SPE Annual Technical Conference and Exhibition, Houston, Texas, USA (September 26–29, 2004).

Fujisawa, G., and Mullins, O.C.: "Live Oil Sample Acquisition and Downhole Fluid Analysis," *Asphaltenes, Heavy Oils, and Petroleomics*, O.C. Mullins, E.Y. Sheu, A. Hammami, and A.G. Marshall (eds.), New York, New York, USA, Springer (2007), 589–616.

Fujisawa, G., Mullins, O.C., Dong, C., Carnegie, A., Betancourt, S.S., Terabayashi, T., Yoshida, S., Jaramillo, A.R., and Haggag, M.: "Analyzing Reservoir Fluid Composition In-Situ in Real Time: Case Study in a Carbonate Reservoir," paper SPE 84092 presented at the SPE Annual Technical Conference and Exhibition, Denver, Colorado, USA (October 5–8, 2003).

Fujisawa, G., Mullins, O.C., Terabayashi, T., Jenet, F.A., van Agthoven, M.A., and Rabbito, P.A.: "Methods and Apparatus for Determining Chemical Composition of Reservoir Fluids," US Patent No. 7,095,012 (August 22, 2006).

Fujisawa, G., van Agthoven, M.A., Jenet, F., Rabbito, P.A., and Mullins, O.C.: "Near-Infrared Compositional Analysis of Gas and Condensate Reservoir Fluids at Elevated Pressures and Temperatures," *Applied Spectroscopy* (December 2002) 56, No. 12, 1615–1620.

George, G.N., and Gorbaty, M.L.: "Sulfur K-Edge X-Ray Absorption Spectroscopy of Petroleum Asphaltenes and Model Compounds," *Journal of the American Chemical Society* (1989) 111, 3182–3186.

Groenzin, H., and Mullins, O.C.: "Molecular Size and Structure of Asphaltenes from Various Sources," *Energy & Fuels* (2000) 14, No. 3, 677–684.

Groenzin, H., and Mullins, O.C.: "Asphaltene Molecular Size and Structure," *Journal of Physical Chemistry A* (1999) 103, No. 50, 11237–11245.

Groenzin, H., Mullins, O.C., and Mullins, W.W.: "Resonant Fluorescence Quenching of Aromatic Hydrocarbons by Carbon Disulfide," *Journal of Physical Chemistry A* (1999), No. 11, 1504–1508.

Hammami, A., Phelps, C.H., Monger-McClure, T., and Little, T.M.: "Asphaltene Precipitation from Live Oils: An Experimental Investigation of Onset Conditions and Reversibility," *Energy & Fuels* (2000) 14, No. 1, 14–18.

Hammami, A., and Ratulowski, J.: "Precipitation and Deposition of Asphaltenes in Production Systems: A Flow Assurance Overview," *Asphaltenes, Heavy Oils, and Petroleomics*, O.C. Mullins, E.Y. Sheu, A. Hammami, and A.G. Marshall (eds.), New York, New York, USA, Springer (2007), 617–660.

Hammond, P.S.: "One- and Two-Phase Flow During Fluid Sampling by a Wireline Tool," *Transport in Porous Media* (June 1991) 6, No. 3, 299–330.

Hashem, M.N., Elshahawi, H., and Ugueto, G.: "A Decade of Formation Testing—Do's and Don'ts and Tricks of the Trade," *Transactions of the SPWLA 45th Annual Logging Symposium*, Noordwijk, The Netherlands (June 6–9, 2004), paper L.

Hashem, M.N., Thomas, E.C., McNeil, R.I., and Mullins, O.C.: "Determination of Producible Hydrocarbon Type and Oil Quality in Wells Drilled with Synthetic Oil-Based Muds," paper SPE 39093 presented at the SPE Annual Technical Conference and Exhibition, San Antonio, Texas, USA (October 5–8, 1997).

Hines, D.R., Wada, N., Garoff, S., Mullins, O.C., Hammond, P., Tarvin, J., Cramer, S.P., and Wiggins, R.: "Method of Analyzing Oil and Water Fractions in a Flow Stream," US Patent No. 5,331,156 (July 19, 1994).

Hodder, M.H., Samir, M., Holm, G., and Segret, G.: "Obtaining Formation Water Chemistry Using a Mud Tracer and Formation Tester in a North Sea Subsea Field Development," paper SPE 88637 presented at the SPE Asia Pacific Oil and Gas Conference and Exhibition, Perth, Australia (October 18–24, 2004).

Hrametz, A.A., Gardner, C.C., Waid, M.C., and Proett, M.A.: "Focused Formation Fluid Sampling Probe," US Patent No. 6,301,959 (October 16, 2001).

Jackson, J.D.: *Classical Electrodynamics* (3rd ed.), New York, New York, USA, John Wiley & Sons, Inc. (1999).

Jenkins, J.A., and White, H.E.: *Fundamentals of Optics* (reprint of 4th ed.), New York, New York, USA, McGraw-Hill Science (2001).

Joshi, N.B., Mullins, O.C., Jamaluddin, A., Creek, J., and McFadden, J.: "Asphaltene Precipitation from Live Crude Oil," *Energy & Fuels* (2001) 15, No. 4, 979–986.

Malinowski, E.R.: *Factor Analysis in Chemistry* (2nd ed.), New York, New York, USA, Wiley-Interscience (1991).

Marshall, A.G., and Rodgers, R.P.: "Petroleomics: The Next Grand Challenge for Chemical Analysis," *Accounts of Chemical Research* (2004) 37, No. 1, 53–59.

McCain, W.D. Jr.: *The Properties of Petroleum Fluids* (2nd ed.), Tulsa, Oklahoma, USA, Penwell Publishing Company (1990).

Mitra-Kirtley, S., Mullins, S.C., van Elp, J., George, S.J., Chen, J., and Cramer, S.P.: "Determination of the Nitrogen Chemical Structures in Petroleum Asphaltenes Using XANES Spectroscopy," *Journal of the American Chemical Society* (1993) 115, 252–258.

Mullins, O.C.: "Method and Apparatus for Determining Gas-Oil Ratio in a Geologic Formation Through the Use of Spectroscopy," US Patent No. 5,939,717 (August 17, 1999).

Mullins, O.C.: "Optical Interrogation of Aromatic Moieties in Crude Oils and Asphaltenes," O.C. Mullins and E.Y. Sheu (eds.), *Structures and Dynamics of Asphaltenes*, New York, New York, USA, Plenum Press (1998), 21–78.

Mullins, O.C.: "Method of Distinguishing Between Crude Oils," US Patent No. 5,266,800 (November 30, 1993).

Mullins, O.C.: "Asphaltenes in Crude Oil: Absorbers and/or Scatterers in the Near-Infrared Region?" *Analytical Chemistry* (1990) 62, No. 5, 508–514.

Mullins, O.C., Beck, G.F., Cribbs, M.E., Terabayashi, T., and Kegasawa, K.: "Downhole Determination of GOR on Single-Phase Fluids by Optical Spectroscopy," *Transactions of the SPWLA 42nd Annual Logging Symposium*, Houston, Texas, USA (June 17–21, 2001), paper M.

Mullins, O.C., Betancourt, S.S., Cribbs, M.E., Dubost, F.X., Creek, J.L., Andrews, A.B., and Venkataramanan, L.: "The Colloidal Structure of Crude Oil and the Structure of Oil Reservoirs," *Energy & Fuels* (2007) 21, No. 5, 2785–2794.

Mullins, O.C., Daigle, T., Crowell, C., Groenzin, H., and Joshi, N.B.: "Gas-Oil Ratio of Live Crude Oils Determined by Near-Infrared Spectroscopy," *Applied Spectroscopy* (February 2001) 55, No. 2, 197–201.

Mullins, O.C., Elshahawi, H., Hashem, M.N., and Fujisawa, G.: "Identification of Vertical Compartmentalization and Compositional Variation by Downhole Fluid Analysis; Towards a Continuous Downhole Fluid Log," *Transactions of the SPWLA 46th Annual Logging Symposium*, New Orleans, Louisiana, USA (June 26–29, 2005), paper K.

Mullins, O.C., Fujisawa, G., Dong, C., Kurkjian, A., Nighswander, J., Terabayashi, T., Yoshida, S., Kinjo, H., and Groenzin, H.: "Determining Dew Precipitation and Onset Pressure in Oilfield Retrograde Condensate," US Patent No. 7,002,142 (February 21, 2006).

Mullins, O.C., Fujisawa, G., Elshahawi, H., and Hashem, M.N.: "Determination of Coarse and Ultra-Fine Scale Compartmentalization by Downhole Fluid Analysis," paper SPE IPTC 10034 presented at the International Petroleum Technology Conference, Doha, Qatar (November 21–23, 2005).

Mullins, O.C., Hines, D.R., Niwa, M., and Safinya, K.: "Apparatus and Method for Detecting the Presence of Gas in a Borehole Flow Stream," US Patent No. 5,167,149 (December 1, 1992).

Mullins, O.C., Hodder, M., Ayan, C., Zhu, Y., and Rabbito, P.: "Method for Validating a Downhole Connate Water Sample," US Patent No. 6,729,400 (May 4, 2004).

Mullins, O.C., Joshi, N.B., Groenzin, H., Daigle, T., Crowell, C., Joseph, M.T., and Jamaluddin, A.: "Linearity of Near-Infrared Spectra of Alkanes," *Applied Spectroscopy* (April 2000) 54, No. 4, 624–629.

Mullins, O.C., Mitra-Kirtley, S., and Zhu, Y.: "The Electronic Absorption Edge of Petroleum," *Applied Spectroscopy* (September 1992) 46, No. 9, 1405–1411.

Mullins, O.C., Rodgers, R.P., Weinheber, P., Klein, G.C., Venkataramanan, L., Andrews, A.B., and Marshall, A.G.: "Oil Reservoir Characterization via Crude Oil Analysis by Downhole Fluid Analysis in Oil Wells with Visible-Near-Infrared Spectroscopy and by Laboratory Analysis with Electrospray Ionization-Fourier Transform Ion Cyclotron Resonance Mass Spectroscopy," *Energy & Fuels* (2006) 21, No. 6, 2448–2456.

Mullins, O.C., Schroeder, R.J., and Rabbito, P.: "Effect of High Pressure on the Optical Detection of Gas by Index-of-Refraction Methods," *Applied Optics* (1994) 33, No. 34, 7963–7970.

Mullins, O.C., and Schroer, J.J.: "Analysis of Downhole OBM-Contaminated Formation Fluid," US Patent No. 6,350,986 (February 26, 2002).

Mullins, O.C., and Schroer, J.: "Real-Time Determination of Filtrate Contamination During Openhole Wireline Sampling by Optical Spectroscopy," paper SPE 63071 presented at the SPE Annual Technical Conference and Exhibition, Dallas, Texas, USA (October 1–4, 2000).

Mullins, O.C., Schroer, J., and Beck, G.F.: "Real-Time Quantification of OBM Filtrate Contamination During Openhole Wireline Sampling by Optical Spectrometry," *Transactions of the SPWLA 41st Annual Logging Symposium*, Dallas, Texas, USA (June 4–7, 2000), paper SS.

Mullins, O.C., Sheu, E.Y., Hammami, A., and Marshall, A.G. (eds.): *Asphaltenes, Heavy Oils, and Petroleomics*, New York, New York, USA, Springer (2007).

Mullins, O.C., Ventura, G.T., Nelson, R.K., Betancourt, S.S., Raghuraman, B., and Reddy, C.M.: "Visible–Near-Infrared Spectroscopy by Downhole Fluid Analysis Coupled with Comprehensive Two-Dimensional Gas Chromatography To Address Oil Reservoir Complexity," *Energy & Fuels* (2008) 22, No. 1, 496–503.

Mullins, O.C., and Zhu, Y.: "First Observation of the Urbach Tail in a Multicomponent Organic System," *Applied Spectroscopy* (February 1992) 46, No. 2, 354–356.

O'Keefe, M., Eriksen, K.O., Williams, S., Stensland, S., and Vasques, R.: "Focused Sampling of Reservoir Fluids Achieves Undetectable Levels of Contamination," paper SPE 101084 presented at the SPE Asia Pacific Oil and Gas Conference and Exhibition, Adelaide, Australia (September 11–13, 2006).

Raghuraman, B., Gustavson, G., Mullins, O.C., and Rabbito, P.: "Spectroscopic pH Measurement for High Temperatures, Pressures and Ionic Strength," *American Institute of Chemical Engineers Journal* (2006) 52, No. 9, 3257–3265.

Raghuraman, B., O'Keefe, M., Eriksen, K.O., Tau, L.A., Vikane, O., Gustavson, G., and Indo, K.: "Real-Time Downhole pH Measurement Using Optical Spectroscopy," paper SPE 93057 presented at the SPE International Symposium on Oilfield Chemistry, The Woodlands, Texas, USA (February 2–4, 2005).

Raghuraman, B., Xian, C., Carnegie, A., Lecerf, B., Stewart, L., Gustavson, G., Abdou, M.K., Hosani, A., Dawoud, A., Mahdi, A., and Ruefer, S.: "Downhole pH Measurement for WBM Contamination Monitoring and Transition Zone Characterization," paper SPE 95785 presented at the SPE Annual Technical Conference and Exhibition, Dallas, Texas, USA (October 9–12, 2005).

Ralston, C.Y., Wu, X., and Mullins, O.C.: "Quantum Yields of Crude Oils," *Applied Spectroscopy* (December 1996) 50, No. 12, 1563–1568.

Reddy, C.M., Nelson, R.K., Sylva, S.P., Xu, L., Peacock, E.A., Raghuraman, B., and Mullins, O.C.: "Identification and Quantification of Alkene-Based Drilling Fluids in Crude Oils by Comprehensive Two-Dimensional Gas Chromatography with Flame Ionization Detection," *Journal of Chromatography A* (April 2007) 1148, No. 1, 100–107.

Rodgers, R.P., and Marshall, A.G.: "Petroleomics: Advanced Characterization of Petroleum-Derived Materials by Fourier Transform Ion Cyclotron Resonance Mass Spectrometry (FT-ICR MS)," O.C. Mullins, E.Y. Sheu, A. Hammami, and A.G. Marshall (eds.), *Asphaltenes, Heavy Oils, and Petroleomics*, New York, New York, USA, Springer (2007), 63–94.

Rodgers, R.P., Tan, X., Ehrmann, B.M., Juyal, P., McKenna, A.M., Purcell, J.M., Schaub, T.M., Gray, M.R., and Marshall, A.G.: "Asphaltene Structure Determined by Mass Spectrometry," presented at the 9th International Conference on Petroleum Phase Behavior and Fouling, Victoria, British Columbia, Canada (June 15–19, 2008), abstract 102.

Ruiz-Morales, Y.: "Molecular Orbital Calculations and Optical Transitions of PAHs and Asphaltenes," E.Y. Sheu, A. Hammami, and A.G. Marshall (eds.), *Asphaltenes, Heavy Oils, and Petroleomics*, New York, New York, USA, Springer (2007), 95–138.

Ruiz-Morales, Y., and Mullins, O.C.: "Polycyclic Aromatic Hydrocarbons of Asphaltenes Analyzed by Molecular Orbital Calculations with Optical Spectroscopy," *Energy & Fuels* (2007) 21, No. 1, 256–265.

Ruiz-Morales, Y., Wu, X., and Mullins, O.C.: "Electronic Absorption Edge of Crude Oils and Asphaltenes Analyzed by Molecular Orbital Calculations with Optical Spectroscopy," *Energy & Fuels* (2007) 21, No. 2, 944–952.

Ryder, A.G.: "Analysis of Crude Oil Petroleum Oils Using Fluorescence Spectroscopy," C.D. Geddes and J.R. Lakowicz (eds.), *Annual Reviews in Fluorescence*, New York, New York, USA, Springer (2005), 169–197.

Safinya, K.A., and Tarvin, J.A.: "Apparatus and Method for Analyzing the Composition of Formation Fluids," US Patent No. 4,994,671 (February 19, 1991).

Sharma, A., Groenzin, H., Tomita, A., and Mullins, O.C.: "Probing Order in Asphaltenes and Aromatic Ring Systems by HRTEM," *Energy & Fuels* (2002) 16, No. 2, 490–496.

Sjöblom, J., Hemmingsen, P.V., and Kallevik, H.: "The Role of Asphaltenes in Stabilizing Water-in-Crude Oil Emulsions," *Asphaltenes, Heavy Oils, and Petroleomics*, O.C. Mullins, E.Y. Sheu, A. Hammami, and A.G. Marshall (eds.), New York, New York, USA, Springer (2007), 549–588.

Tissot, B.P., and Welte, D.H.: *Petroleum Formation and Occurrence* (2nd ed.), Berlin, Germany, Springer-Verlag (1984).

Turro, N.J.: *Modern Molecular Photochemistry*, Menlo Park, California, USA, Benjamin/Cummings Publishing Co., Inc. (1978).

Urbach, F.: "The Long-Wavelength Edge of Photographic Sensitivity and of the Electronic Absorption of Solids," *Physics Review* (1953) 92, No. 5, 1324.

van Agthoven, M.A., Fujisawa, G., Rabbito, P., and Mullins, O.C.: "Near-Infrared Spectral Analysis of Gas Mixtures," *Applied Spectroscopy* (May 2002) 56, No. 5, 593–598.

Venkataramanan, L., Fujisawa, G., Mullins, O.C., Vasques, R.R., and Valero, H.-P.: "Uncertainty Analysis of Visible and Near-Infrared Data of Hydrocarbons," *Applied Spectroscopy* (June 2006) 60, No. 6, 653–662.

Wang, J., and Mullins, O.C.: "Fluorescence of Limestones and Limestone Components," *Applied Spectroscopy* (December 1997) 51, No. 12, 1890–1895.

Wang, X., and Mullins, O.C.: "Fluorescence Lifetime Studies of Crude Oils," *Applied Spectroscopy* (August 1994) 48, No. 8, 977–984.

Weinheber, P., and Vasques, R.: "New Formation Tester Probe Design for Low Contamination Sampling," *Transactions of the SPWLA 47th Annual Logging Symposium*, Veracruz, Mexico (June 4–7, 2006), paper Q.

Xian, C., Raghuraman, B., Carnegie, A., and Goiran, P.-O.: "Downhole pH as a Novel Measurement Tool in Formation Evaluation and Reservoir Monitoring," *Transactions of the SPWLA 48th Annual Logging Symposium*, Austin, Texas, USA (June 3–6, 2007), paper JJ.

Zajac, G.W., Sethi, N.K., Joseph, J.T., Thompson, D.J., and Weiss, P.S.: "Molecular Imaging of Petroleum Asphaltenes by Scanning Tunneling Microscopy: Verification of Structure from 13C and Proton Nuclear Magnetic Resonance Data," *Scanning Microscopy* (1994) 8, No. 3, 463–470.

Zeng, H., Song, Y.-Q., Johnson, D.L., and Mullins, O.C.: "Critical Nanoaggregate Concentration of Asphaltenes by DC-Electrical," *Energy & Fuels* (in press).

Zhu, Y., and Mullins, O.C.: "Temperature Dependence of Fluorescence of Crude Oils and Related Compounds," *Energy & Fuels* (1992) No. 6, 545–522.

Zimmerman, T.H., Pop, J.J., and Perkins, J.L.: "Down Hole Method for Determination of Formation Properties," US Patent No. 4,936,139 (June 26, 1990).

Zimmerman, T.H., Pop, J.J., and Perkins, J.L.: "Down Hole Tool for Determination of Formation Properties," US Patent No. 4,860,581 (September 23, 1989).

Index

A

absorption of light, 92
acoustic frequency, 117
acoustic velocity measurement, 14
Africa, venting of carbon dioxide from Lake Nyos, 33, *33*
aggregation number, 143
alkanes
  biodegradation of, 25
  CFA analysis, 125, *126*
  compositional analysis of, 123
  ionization by ESI, 163
  in oilfield waxes, 101
analytical chemistry, 84–86
angle of refraction, 130
antennas, 131
Archimedes buoyancy, 22
argon, molecular polarizability of, 133, *134*
asphaltenes, 36, 82
  aggregation threshold for, 13
  centrifugation studies of, 20
  chromophores in, 112, *112*
  color of, 100
  coloration of, 18–19, *18*, 22, 113
  compartments and, 50, 55
  crude oil variations, 30
  deposition of, 56
  destabilization of, 55, 114
  diffusion constants of, 20
  electronic absorption edge, 107
  equilibrium fluid distribution of, 15–24
  flocculation of, 96, *98*, 99, 100, 156
  flow assurance, 55
  gas chromatography, 164
  isolating, 114
  molecular structure of, 104
  nanoaggregates of, 20–21, *21*, 23, 160
  as nanocolloids, 94, 143
  onset pressure, 99, *99*
  plugging by, 56, *57*
  quenching in, 143
  in toluene, 20, 21, *21*
  viscosity variation with asphaltene content, 26, 27
  X-ray spectroscopy and, 88
Athabasca bitumen, 37, 72
atomic polarizability, 133

B

basin modeling, 38, 62
Beer-Lambert law, 92
benzene, 102, *103*, 104, 107–108
Betancourt, S.S., 22
biodegradation
  of hydrocarbons, 25–26, 25
  viscosity variation with, 26
biogenic methane, 28, *28*, 29, 30, 37
biomarkers
  analysis using, 4, 26, 42, *42*
  kerogen catagenesis, 41, 42, *42*, 107
bitumen, 35, 37
black oils, 15, 24, 55, 143
blackbody radiation, 89
blue shift, 146
borehole temperature, 71
borehole wall, fluorescence log of, 139–140, *140*, 141
Brewster angle, 131
bubblepoint, 129
bubblepoint line, 129, *129*
bulk spectroscopy, 85

C

Cantor, Georg, 51
carbon, isotope ratios of, 28–29, 37
carbon dating, 28
carbon dioxide
  NIR spectra of, 123–124, *124*
  reservoir charging, 32–34
  venting from Lake Nyos, 33, *33*
carbon isotope ratios, 28–29, 37
Carnegie, A.J., 11, 59
catacondensed status, 108
catagenesis, of kerogens, 37, 41, 42, *42*, 107
CC ring stretch mode, 104
centrifugation studies
  of asphaltenes, 20
  of hydrocarbons, 7–9, *7–9*
CFA* Composition Fluid Analyzer, 3, 10, 32, 55, 120, 124, 125
chain of custody, 156–157, *158*
chemical analysis, 84–86
  See also downhole fluid analysis
chemical bonds, 102, 118
chemical waste disposal, chain of custody in, 156
Chicago PDB Marine Carbonate Standard (CPDB), 29
China, gas composition from Hainan Island reservoir, 34, *34*
chromatography, 84
chromophores
  in asphaltenes, 112, *112*
  in crude oil, 87, 100–115, 142, 143
  Energy Gap law and, 152
Clifford, P.J., 16
CMR* Combinable Magnetic Resonance log, 140, *141*
CNAC. *See* critical nanoaggregate concentration
coalbed methane, 36
collisional quenching, 147, 149
collisional relaxation processes, 146
colloidal systems, light scattering of, 93, *93*
color, of crude oil, 110–115
coloration
  of asphaltenes, 18–19, *18*, 22, 113
  of crude oils, 74, 104, *105*, 113, 114, 115

Index   177

of OBM filtrates, 75
   temperature dependence of, 113
compartmentalization, 43–52, 55
connate water, 153
contamination. *See* sample contamination
continuous downhole fluid log, 158
continuous gas phase, 136
CPDB. *See* Chicago PDB Marine Carbonate Standard
Cracking, 37
Creek, Jefferson, 16
Cribbs, Myrt E. (Bo), 18
critical angle, 131, 133
critical nanoaggregate concentration (CNAC), 20
crude oil
   analytical chemistry, 84–86
   biodegradation of, 25
   bubblepoint line, 129, *129*
   chromophores in, 87, 110–115, 142, 143
   classification of, 83t
   coloration of, 74, 104, *105*, 113, 114, 115
   electronic band-gap transition energies, 150
   emulsification of, 139
   Energy Gap law, 149, *149*, 150–151
   fluorescence emission spectra of, 139, *139*, *144–145*
   fluorescence of, 135–152
   fluorophores in, 110, 142, 143, 151
   GOR of, 120, *121*
   intersystem crossing, 151
   NIR overtones of, 121, *122*
   phase transitions in, 55, 83
   quantum yield of, 150, *150*, 152, *152*
   separation science, 85
   spectral analysis of, 92
   spider diagrams to compare samples, 165, *165*
   Vis-NIR spectra of, 74, 75, 105, 106, 108
   visible color of, 87, 110–115

crude oil fluorescence
   downhole fluorescence, 135–140, *141*
   science of, 142–152

D

Dake, Laurence P., 43, 60
Davies, T. Harrison, 29
DBE. *See* double bond equivalent
dead oils, compositional gradient, 3, 4
deemulsification, 137, *138*
deepwater workflow, 61
delocalization, 102, 104
density, 12
dewpoint line, 135
DFA. *See* downhole fluid analysis
diffusion constant, 143
dispersion, of colloidal systems, 93
dissociation constant, 154
Dong, C., 28, 50
double bond equivalent (DBE), 162, *163*
downhole bottles, 156
downhole fluid analysis (DFA)
   about, 1–2, 42, 84
   assumptions regarding, 166
   chain of custody studies using, 157, *158*
   compartmentalization, 43–52, 55
   equilibrium distribution of asphaltenes, 15–24
   equilibrium distribution of hydrocarbons, 10–14
   errors addressed by, 12
   fluid color predictions by, 22
   fluid density measurements by, 14–15
   GC×GC and, 164
   GOR, 8–11, 15, 16, 22, 28, 30, *31*, 39–40, 42, 50-52, 75, 76–77, 82
   nonequilibrium distribution of hydrocarbons, 24–34
   optics tools, 131
   photoexcitation and, 86, 118, 142
   production and miscible injection, 55–56, 57

quasi-continuous downhole fluid log, 158–159
recommended workflow, 61–62
reservoir charge history, 35–43
senior technologists for, 62
uses of, 158
visible-near-infrared (Vis-NIR) spectroscopy, 72–82
workflow flaws, 60–61
downhole fluorescence, 135–140, *141*
downhole sample acquisition, spectroscopy for, 85
Dubost, F.X., 11, 22
dynamic model of reservoir, 60

E

earthquakes, 46, *46*
Eddington, A.S., 60
Einstein, Albert, 60, 101
electromagnetic radiation, 87–88, *88*
electronic absorption edge, 107
electronic orbitals, 101–102, *102*
electronic transitions, 87
electrophoresis, 84
electrospray ionization Fourier transform ion cyclotron resonance mass spectroscopy (ESI FT-ICR MS), 160–161, *161*
Elshahawi, H., 28, 50, 104
emulsions, 137, *138*
energies of vibrations, 117–118
Energy Gap law, 149, *149*, 150–151
England, W.A., 16
EOS modeling, 11, *11*, 12
equilibrium distribution
   of asphaltenes, 15–24
   of hydrocarbons, 10–14
Eriksen, Kåre Otto, 10, 58
errors, addressed for DFA, 12
ESI FT-ICR MS. *See* electrospray ionization Fourier transform ion cyclotron resonance mass spectroscopy

Euler's number, 92
extraheavy oil, 51

F

faults, geophysical scaling, 49, *49*
Fenn, John, 163
Fermi edge, 107
fiber optics, 89, 131
FID. *See* flame ionization detector
flame ionization detector (FID), 164
flocculation, of asphaltenes, 96, *98*, 99, 100, 156
flow assurance, asphaltenes, 55
flow communication, 45, 50
FLT. *See* fluorescence logging tool
fluid density inversions, 50, 51, *51*, 52, 54
fluid density measurements, by downhole fluid analysis, 14–15
fluid identification
  downhole Vis-NIR measurements,121
  GOR, 121
  by Vis-NIR spectroscopy, 74-75, 75, 76
Fluid Profiling® characterization and quantification, 1, 2
fluorescence, 135
  blue shift, 146
  of crude oils, 135–152
  decay curves, 146, 147, *147*
  downhole fluorescence, 135–140, *141*
  of emulsions, 137, *138*
  lifetime reduction, 147
  measurement of, 86, 142–152
  quenching, 143, 146, 147, 149, 152
  red shift, 142, 143, 152
  retrograde dew transition, 136 137, *136*
  Rydberg equation, 151
fluorescence blue shift, 146
fluorescence decay curves, 146, 147, 147

fluorescence emission spectra, of crude oils, 139, *139*, *144–145*
fluorescence log, of borehole wall, 139–140, *140*, *141*
fluorescence logging tool (FLT), 140, *140*, *141*
fluorescence red shift, 142, 143, 152
fluorophores, in crude oil, 110, 142, 143, 151
forbidden bands, 88–89, 118
formation water samples, pH measurement of, 153
Franck-Condon factor, 104
Fuex, A.N., 7
fused aromatic rings, in PAHs, 108, *109–111*, 112
fused ring structure, 104

G

gamma rays, 88, *88*
gas
  compositional grading of, 6
  continuous gas phase, 136
  critical angle of, 133
  detection of, 130–131, *130*, *132*
  index of refraction of, 130–133, *134*
  phase transition to, 129–134, 156
  separator gas injection, 56
  under borehole pressure, 132
gas chromatography (GC), 84
gas-prone kerogens, 35
gas window, of kerogen, 35
gases. *See* gas
GC. *See* gas chromatography
GC×GC, 164
GC×GC-FID chromatogram, *164*
geophysical objects, 48, *48*
Gisolf, Adriaan, 12
GOR
  calculation of, 122
  color and, 9, 22
  of crude oil, 120, *121*
  density inversion measurement

and, 51–52
  downhole measurement of, 75, 76–77
  for extraheavy oils, 51
  gradients, 8, 11
  importance of, 82
  interpretation algorithm for, 121
  units of, 120
  variations of, 15, 16, 28, 30, *31*, 39–40
Green River kerogen, 35

H

half alkane carbon, 36
harmonic oscillator, 116, 118
Hashem, Mohamed, 47
heavy oil, 55
  nanoaggregates of, 24
  NIR spectra of, 137, *138*
  petroleome, 160
heavy oil–water invert emulsions, 139
helium, molecular polarizability of, 133, *134*
n-heptane, Vis-NIR spectra of, 118, 119, *119*, *120*
Herschel, William, 74
heteroatom, 162
HI. *See* hydrogen index
high overtones, 118
high-pressure gas detector cell, *132*, 133, *134*
high-Q ultrasonic measurements, 12, 13
HO-LU gap, 150, 151
Høier, L., 5
Høier-Whitson gradients, 5, 7, 11, 15, 24
Hooke's law, 116
hopane, 26
Hows, M., 28
hydrocarbons, *82*
  biodegradation of, 25–26, 25
  CFA analysis of, 125, *126*

Index 179

chain of custody of samples of, 156–157
chemical behavior of, 92
compositional analysis of, 116–128
DFA and, 10–14
effect of pressure on, 127, *127*, *128*
electron orbitals in, 101, *102*
equilibrium distribution of, 5–14
fluid density measurements, 15
nonequilibrium distribution of, 24–34
PAHs, 103, *103*, 106–108, *109–111*, 112
petroleomics, 85, 159–166
phase transition to dewpoint line, 135
phase transition to gas, 129–134
reservoir charging, 28–30, *28*, *29*, *31*, *32*
hydrogen index (HI), 20
hydrogen sulfide, kerogens and, 36

I

IFA* Insitu Fluid Analyzer, *73*, 120
incandescent lamps, 89
index of refraction, for gas detection, 130–133, *134*
infinity, 51
infrared-emitting diode (IRED), *132*
InGaAs photodiode detectors, 89–90
InSitu pH* Reservoir Fluid pH Sensor, 3
interfacial surface area, in colloids, 93
intersystem crossing, 151
invert emulsions, 139
IRED. *See* infrared-emitting diode

K

kerogens, 35–37, 52
catagenesis of, 37, 41, 42, *42*, 107
cooking time of, 37
defined, 35
gas window of, 36
heating rate of, 37
hydrogen sulfide and, 36
oil window of, 36, 53
types of, 35–36, *36*

Khong, K.C., 34
Kimmeridge clay kerogen, 37, *38*

L

Lake Nyos, carbon dioxide venting from, 33, *33*
Larter, S., 4
Le Chatelier's principle, 5, 6
LFA* Live Fluid Analyzer, 3, 96, 120,123
Libby, W.F., 29
light absorption, 90, *90*
light polarization, 130–132, *131*, *132*
light scattering, 92, 93–100, *95*, *97*, 132
linear polarization, 130
litmus paper, 153
long-wavelength waves, *95*
low overtones, 118

M

macerals, 35
maltene, 26, 27
mass density, 127, *127*, *128*
mass spectroscopy, 84, 160, 163
matrix, 124
McKinney, Dan, 28
MDT* Modular Formation Dynamics Tester, 2, 10–11, *10*, 22, 56, 72, *73*, 77, *78*, 85
mechanical oscillator, 117
methane, 36
biogenic, 28, *28*, 29, 30, 37
CFA analysis, 125, *126*
compositional analysis of, 123
critical angle of, 133
isotope ratios, 28–29, 37
molecular polarizability of, 133, *134*
NIR spectra of, 123–124, *124*
sigma (σ)-bonds of, 101, *102*
thermogenic, 28, 37
Vis-NIR spectra of, 119, *120*
micelle formation, 13
microwave ovens, 89

microwave radiation, 89
mid-infrared (MIR) spectral range, 88, *88*, 89, 90, 118
miscible injection, 55–56, 57
Mitchell, Brett, 10
molar absorptivity, 92
molecular energy levels, 86–90, *88*
molecular excitation processes, 86
molecular polarizability, 133
molecular spectroscopy, 86–100
molecular transitions, 87
molecular vibration, 116
mud filtrate, 71–72, 96
Muggeridge, A.H., 16
Mullins, O.C., 11, 22

N

nanoaggregates
of asphaltene, 20–21, *21*, 23, 160
of heavy oil, 24
nanocolloids, in asphaltenes, 94, 143
naphthalene, 103, *103*
near-critical fluids, 5, *5*, 6, *6*
near-infrared (NIR) spectral range, 88–89, 90, 118
near-infrared (NIR) spectroscopy, 15, 74, 123–124, *124*, 137, *138*
near-saturation fluids, 5, 5
neutron flux, 29
NIR spectroscopy. *See* near-infrared (NIR) spectroscopy
nitrogen, molecular polarizability of, 133, *134*
nonequilibrium distribution, of hydrcarbons, 24–34
nonhydrocarbons, reservoir charging, 32–34, *33*, *34*

O

OBM. *See* oil-base mud
OCM* algorithm, 79

OD. *See* optical density
OFA* Optical Fluid Analyzer, 96
offshore projects, 43
oil-base mud (OBM), purposes of, 71
oil-base mud (OBM) contamination, 52
oil-base mud (OBM) filtrate
    asphaltene stability and, 114
    coloration of, 75
    differentiation of crude oil from downhole, 100
    Vis-NIR spectra of, 74, 75
oil chemistry, 82–83, 83–84, *83*
    analytical chemistry, 84–86
    molecular spectroscopy, 86–100
oil-prone kerogens, 35
oil sample acquisition. *See* sample acquisition
oil window, of kerogen, 36
oilfield waxes, 101
O'Keefe, Michael, 10, 58
onset pressure, asphaltenes, 99, *99*
onshore reservoirs, 43
openhole samples
    acquisition of, 71
    chain of custody in, 156–157, *158*
optical channels, GOR, 121, 123
optical density (OD), 74, 93
optical light scattering, 92
optical principles, spectroscopy, 90–93
optical spectroscopy, 86
optical transmission, 91
organic scale, 139
oscillators, 116, 131

P

p-polarization, 130, *130*, *132*
PAHs. *See* polycyclic aromatic hydrocarbons
Pal-Rhodes viscosity model, 26
partial least square (PLS) technique, 124

PCA. *See* principal components analysis
PCR. *See* principal components regression
pericondensed status, 108
petroleome, 160
petroleomics, 85, 159–166
Pettersen, Rolf Magne, 12
pH
    calculation of, 154
    colorimetric analysis of, 153, 154, *155*
    downhole measurement of, 153 155, *155*, *156*
    water chemistry and, 58, *59*
pH dye, 153, *155*
phase behavior, 55
phase transitions
    of crude oil, 129–134, 135
    retrograde dew, 11, 74, 136–137, *136*, 156
photodiode detectors, 89–90
photoexcitation, 86, 118, 142
phytane, 26
pi(π)-bonds, 102
pi(π)-electrons, 102, 104
PLS technique. See partial least square (PLS) technique
polarization, 130–132, *131*, *132*
Polaroid® sunglass lenses, 131
polycyclic aromatic hydrocarbons (PAHs), 103, *103*, 106, 107
    absorption band spectral locations, 108, *109–111*, 112
    in fused aromatic rings, 108, *109–111*, 112
potential energy curves, 116, *117*
predictive science, 160
predictive workflow, 61
pressure communication, 45, 50
pressure gradient measurements, 12
pressure gradients, 14-15
principal components analysis (PCA), 124

principal components regression (PCR), 124
pyrolitic reactions, 37

Q

Q factor (quality factor), 12
quantum mechanics, crude oil coloration and DFA, 110–115
quantum yield, of crude oil, 150, *150*, 152, *152*
quasi-continuous downhole fluid log, 158–159
quencher, 143
quenching, 146, 152
    in asphaltenes, 143
    collisional quenching, 147, 149
Quicksilver Probe* focused fluid extraction, *73*, 81, *81*, 114–115, *115*

R

radioactive decay, 91
Raghuraman, B., 58
Ratulowski, J., 7
Rayleigh scattering, 94
RDCs. *See* Reservoir Domain Champions
red shift, 142, 143, 152
regression vector, 125
reservoir, models of, 60
reservoir charge history, 35–43
reservoir charging
    carbon dioxide, 32–34
    hydrocarbons, 28–30, *28*, *29*, *31*, *32*
    nonhydrocarbons, 32–34, *33*, *34*
Reservoir Domain Champions (RDCs), 62
reservoir fluid analysis
    by centrifuging, 7–9, *7–9*
    equilibrium distribution, 10–24
    nonequilibrium gradients, 24–34
    prior to DFA, 2

reservoir fluids, 2
    complexity of, 3–4
    equilibrium and, 38–39, *38*
    thermal gradients in, 24
reservoir rock evaluation, 4
resonator, Q (quality) factor, 12
retrograde condensation, 135–136
retrograde dew, 11, 74, 136–137, *136*, 156
retrograde one-phase dew, 11
ring stretch mode, 104
ring system geometry, 108
rotational transitions, 89
Rydberg constant, 151
Rydberg equation, 151

S

s-polarization, 131
sample acquisition, 71
    cost of, 158
    oil chemistry and, 82–83, 83–84, *83*
    See also downhole sample acquisition; openhole samples
sample bottles, 156
sample contamination
    chain of custody and, 156–157
    laboratory methods to quantify, 81–82
    monitoring of, 75, 76–77, *76–78*, 79, *80*, 81–82, 155
    oil-base mud (OBM) contamination, 52
sapphire windows, 89, 131, *132*
SARA analysis, 84–85
SARA fractions, 142
saturated hydrocarbons, electron orbitals in, 101, *102*
scalars, 124
scattering, 93–100
"Schlumberger blue" dye colorant, 155
scientific workflow, 60
SDS. *See* sodium dodecyl sulfate

sedimentation equation, 96
semiconductors, 107
separation science, 84, 166
    See also downhole fluid analysis
separator gas injection, 56
short-wavelength waves, *95*
sigma (σ)-bonds, 101, *102*
silica, 89
"skimming technique," 81
Snell's law, 130
sodium dodecyl sulfate (SDS), 13, *14*
sound, speed of, 117
spectral analysis, 85, 86, 92, 123, 166
spectrometers, 90
spectroscopy, 85, 90–93
speed of sound, 117
spider diagram, 165, *165*
Stainforth model, 40, *41*
static model of reservoir, 60
steady states, 24
Stern-Volmer plot, 148, *148*
Stokes effect, 142
Stokes viscous drag coefficient, 96
surfactants, 13–14, 139

T

Tahiti field, asphaltenes in, 15–16, *16–19*, 18, 22
2,4,10,14-tetramethylhexadecane, 26
tetraorganic acids, 139
thermal gradients, in reservoir fluids, 24
thermogenic methane, 28, 37
toluene, asphaltenes in, 20, 21, *21*
total absorbance, 92
total internal reflection, 131
transition zones, 58, 153–155
Type I kerogen, 35, *36*
Type II kerogen, 35, *36*

Type III kerogen, 35–36, *36*
Type IV kerogen, 36
Tyrihans reservoir, 10–11, *10*, 12

U

Urbach equation, 107, 123
Urbach slope, 107
Urbach tail, 107

V

van Krevelen diagram, 35, 36
vectors, 124
vibrational frequency, 117
vibrational levels, quantization of, 116, *117*
vibrational modes, 117, 118
viscosity, variation with biodegradation, 26
viscous drag force, 96
visible-near-infrared (Vis-NIR) spectral range, 89
visible-near-infrared (Vis-NIR) spectroscopy, 72–82, 86
    of crude oils, 74, 75, 105, 106, 108
    for fluid identification, 74–75, 75, 76, 105, 106
    of n-heptane, 118, 119, *119*, *120*
visible spectral range, 87–88, *88*
vitrinite, 37

W

waste disposal. *See* chemical waste disposal
water
    absorption spectrum of, 128, *128*
    Vis-NIR spectra of, 74, *75*
water-base mud (WBM), 71, 140
water-base mud (WBM) filtrate, 58, 153, 155
water-in-oil emulsions, 139

waxes, 101, 156
WBM. *See* water-base mud
Wein's displacement law, 89
Westrich, J.T., 7
Whitson, C.H., 5
Wilhelms, A., 4
Williams, Steve, 12
wireline tool spectrometers, 125

X

X-ray radiation, 88, *88*
X-ray Raman spectroscopy, 108
X-ray spectroscopy, 88
Xian, C., 58

Z

Zumberge, J., 29
Zuo, Julian, 12

Other Books Co-edited by the Author

Asphaltenes, Heavy Oils, and Petroleomics, O.C. Mullins, E.Y. Sheu, A. Hammami, and A.G. Marshall (eds.), New York, New York, USA, Springer (2007).

Structures and Dynamics of Asphaltenes, O.C. Mullins and E.Y. Sheu (eds.), New York, New York, USA, Plenum Press (1998).

Asphaltenes: Fundamentals and Applications, E.Y. Sheu and O.C. Mullins (eds.), New York, New York, USA, Plenum Press (1995).